Springer Theses

Recognizing Outstanding Ph.D. Research

For further volumes:
http://www.springer.com/series/8790

Aims and Scope

The series "Springer Theses" brings together a selection of the very best Ph.D. theses from around the world and across the physical sciences. Nominated and endorsed by two recognized specialists, each published volume has been selected for its scientific excellence and the high impact of its contents for the pertinent field of research. For greater accessibility to non-specialists, the published versions include an extended introduction, as well as a foreword by the student's supervisor explaining the special relevance of the work for the field. As a whole, the series will provide a valuable resource both for newcomers to the research fields described, and for other scientists seeking detailed background information on special questions. Finally, it provides an accredited documentation of the valuable contributions made by today's younger generation of scientists.

Theses are accepted into the series by invited nomination only and must fulfill all of the following criteria

- They must be written in good English.
- The topic should fall within the confines of Chemistry, Physics and related interdisciplinary fields such as Materials, Nanoscience, Chemical Engineering, Complex Systems and Biophysics.
- The work reported in the thesis must represent a significant scientific advance.
- If the thesis includes previously published material, permission to reproduce this must be gained from the respective copyright holder.
- They must have been examined and passed during the 12 months prior to nomination.
- Each thesis should include a foreword by the supervisor outlining the significance of its content.
- The theses should have a clearly defined structure including an introduction accessible to scientists not expert in that particular field.

Matthias J. N. Junk

Assessing the Functional Structure of Molecular Transporters by EPR Spectroscopy

Doctoral Thesis accepted by
Johannes Gutenberg-Universität Mainz,
Mainz, Germany

Author
Dr. Matthias J. N. Junk
Max Planck Institute
for Polymer Research
Mainz, Germany

Supervisor
Prof. Hans Wolfgang Spiess
Max Planck Institute
for Polymer Research
Ackermannweg 10
55128 Mainz
Germany
e-mail: spiess@mpip-mainz.mpg.de

ISSN 2190-5053
ISBN 978-3-642-25134-4
DOI 10.1007/978-3-642-25135-1
Springer Heidelberg Dordrecht London New York

e-ISSN 2190-5061
e-ISBN 978-3-642-25135-1

Library of Congress Control Number: 2011942407

Printed on acid-free paper

Springer is part of Springer Science+Business Media (www.springer.com)

Parts of this thesis have been published in the following journal articles:

M. J. N. Junk, H. W. Spiess, D. Hinderberger, The Distribution of Fatty Acids Reveals the Functional Structure of Human Serum Albumin, *Angew. Chem.* **2010**, *122*, 8937–8941, *Angew. Chem. Int. Ed.* **2010**, *49*, 8755–8759.
(*Chapter 3.2*)

M. J. N. Junk, H. W. Spiess, D. Hinderberger, DEER in Biological Multispin-Systems: A Case Study on the Fatty Acid Binding to Human Serum Albumin, *J. Magn. Reson.* **2011**, *210*, 210–217.
(*Chapter 3.3*)

M. J. N. Junk, H. W. Spiess, D. Hinderberger, Characterization of the Solution Structure of Human Serum Albumin Loaded with a Metal Porphyrin and Fatty Acids, *Biophys. J.* **2011**, *100*, 2293–2301.
(*Chapter 3.4*)

J. Zhang, J. Luo, X. X. Zhu, M. J. N. Junk, D. Hinderberger, Molecular Pockets Derived from Cholic Acid as Chemosensors for Metal Ions, *Langmuir* **2010**, *26*, 2958–2962.
(*Chapter 4.2*)

J. Zhang, M. J. N. Junk, J. Luo, D. Hinderberger, X. X. Zhu, 1,2,3-Triazole-containing Molecular Pockets Derived from Cholic Acid: The Influence of Structure on Host–Guest Coordination Properties, *Langmuir* **2010**, *26*, 13415–13421.
(*Chapter 4.3*)

M. J. N. Junk, U. Jonas, D. Hinderberger, EPR Spectroscopy Reveals Nano-Inhomogeneities in the Structure and Reactivity of Thermoresponsive Hydrogels, *Small* **2008**, *4*, 1485–1493.
(*Chapter 5*)

B. C. Dollmann, M. J. N. Junk, M. Drechsler, H. W. Spiess, D. Hinderberger, K. Münnemann, Thermoresponsive, Spin-labeled Hydrogels as Separable DNP Polarizing Agents, *Phys. Chem. Chem. Phys.* **2010**, *12*, 5879–5882.
(*Chapter 6*)

M. J. N. Junk, W. Li, A. D. Schlüter, G. Wegner, H. W. Spiess, A. Zhang, D. Hinderberger, EPR Spectroscopic Characterization of Local Nanoscopic Heterogeneities during the Thermal Collapse of Thermoresponsive Dendronized Polymers, *Angew. Chem.* **2010**, *122*, 5818–5823, *Angew. Chem. Int. Ed.* **2010**, *49*, 5683–5687.
(*Chapter 7.2*)

M. J. N. Junk, W. Li, A. D. Schlüter, G. Wegner, H. W. Spiess, A. Zhang, D. Hinderberger, EPR Spectroscopy Provides a Molecular View on Thermoresponsive Dendronized Polymers Below the Critical Temperature, *Macromol. Chem. Phys.* **2011**, *212*, 1229–1235.
(*Chapter 7.3*)

M. J. N. Junk, W. Li, A. D. Schlüter, G. Wegner, H. W. Spiess, A. Zhang, D. Hinderberger, Formation of a Mesoscopic Skin Barrier in Mesoglobules of Thermoresponsive Polymers, *J. Am. Chem. Soc.* **2011**, *133*, 10832–10838.
(*Chapter 7.4*)

Supervisors' Foreword

Molecular transporters are of seminal importance in modern materials science and in the life sciences (biology, pharmacy, medicine) alike. Materials science-oriented chemistry is heavily focused on rather simple synthetic or biomimetic systems for the directed transport and delivery of small molecule drugs, while the biology-oriented view is more focused on much more complex biomacromolecules. Interestingly, the prevalent interactions between the—often amphiphilic—transported molecules and the macromolecular carriers are among the least understood aspects of such systems.

In his doctoral thesis, Matthias Junk opens up remarkably new paths to studying and characterizing such interactions between transported materials and carriers. The remarkable versatility of his approach is evident from the large variety of systems that he studied on the molecular or nanometer scale with different techniques of electron paramagnetic resonance (EPR) spectroscopy on probe molecules. By introducing stable radicals into the respective systems, he can directly follow the transport processes, through the interactions between transporter and transported molecule. His EPR-spectroscopic techniques range from simple continuous wave (CW) measurements of EPR spectra, which can be recorded with a desk-top-spectrometer, to state-of-the-art measurements of distances between spin labels on the nm-range by pulsed double electron–electron resonance (DEER) in high-end equipment.

Especially noteworthy are his studies of the structure of human serum albumin, the most abundant protein in human blood plasma, which serves as a transporting agent for various endogenous compounds and drug molecules. Notably the structure of the loaded protein he finds in solution significantly differs from that known from X-ray crystallography. Beyond this remarkable finding itself, Matthias Junk has established his approach of measuring distances between transported molecules to gain deep insights into the solution structure of functional proteins. This is now used to characterize the effect of interaction with different solvents, in particular ionic liquids, on the protein's solution structure.

Another highlight of the thesis concerns probing the collapse of thermoresponsive hydrogels from the entrapped molecules' point of view, which delivered new insights into the nanoscale heterogeneities that arise during these thermal transitions. He also used hydrogels for removing the polarizing spin labels in the emerging field of enhancing NMR signals by dynamic nuclear polarization (DNP). Removal of these radicals is crucial for the foreseen applications of this method in clinical magnetic resonance imaging (MRI).

Matthias Junk's thesis is a fine example of how magnetic resonance spectroscopy, carried out in great depth and deep understanding, can yield unique and highly specific information about many different chemical and biological systems. Such an approach pushes the boundaries even for rather "mature" techniques like EPR spectroscopy and delivers remarkable new insights into materials.

Mainz, September 2011

Dariush Hinderberger
Hans Wolfgang Spiess

Preface

When I first came into contact with EPR spectroscopy during my diploma thesis, I was skeptical. How should radicals be of any help in the structural determination of complex materials? After all, I was merely interested in the radicals themselves but in the environment surrounding them. I forgot that unlike nuclear spins, electrons experience a large spatial distribution enabling them to 'see' and 'sense' far into their neighborhood. Although other characterization methods I used at that time also provided quite insightful results, it was EPR spectroscopy that delivered an insight into the nanoscopic structural and dynamic arrangement of molecules. This triggered my decision to solely focus on EPR spectroscopy as a tool to probe complex macromolecular systems in my PhD research.

The unique sensing capabilities of electron spins have been known for a long time and led many researchers to exploit radicals as spectral sensors. The concept of spin probing and spin labeling has come of age and has been applied to many biological and synthetic systems. Yet, the recent advent and continuing development of pulse techniques keep expanding the potential and applicability of EPR spectroscopy. Although still a scientific niche, it is a fast growing field that nicely complements established scattering and spectroscopic techniques. Particularly, the possibility to reliably infer distances in the range of several nanometers between two spin labels fueled the expansion of EPR spectroscopy in the field of structural and molecular biology. Such distance constraints can be used to determine the 3D structure and assembly of macromolecular systems like proteins and nucleic acids. However, the synthetic effort for obtaining and labeling specific protein mutants is high and tedious. Moreover, only one distance constraint can be obtained by one spin labeled mutant and distance measurement in most cases. Focusing on transporting materials this triggered the idea to approach the problem differently. Instead of labeling the fatty acid transporter human serum albumin itself, I simply replaced the transported guests by EPR-active fatty acids. The insights into the functional structure of albumin from the viewpoint of the transported guest molecules and the implications of such self-assembled systems for the respective EPR method are presented in Chap. 3. Yet, one does not always require state-of-the-art technology to obtain novel and exciting findings. This is demonstrated in Chaps. 5 and 7, where the thermal collapse of thermoresponsive polymeric systems is studied by one of the most elementary nitroxide spin probes and simple continuous wave EPR spectroscopy.

The broad applicability of modern EPR spectroscopy in structural biology and materials science is emphasized by the examined systems: With a transport protein, a biomimetic host system, and responsive hydrogels and dendronized polymers, three substantially different transporting agents were examined by a variety of EPR spectroscopic methods and probes. In either case, EPR spectroscopy delivered illuminating and sometimes surprising insights into the molecular transporters, leading to a detailed molecular understanding of the materials' functions.

This thesis would not have been possible without the support of many people. First of all, I wish to thank my supervisors Prof. Dr. Hans Wolfgang Spiess and Dr. Dariush Hinderberger for their continuing interest in my research, their excellent ideas and their comprehensive support. Many people answered my never ending questions. Representative for all these people, I want to thank Dr. Uli Jonas for introducing me into the scientific world and Prof. Dr. Gunnar Jeschke for many helpful suggestions related to orientation selective and flip angle dependent DEER. I thank my co-workers from the University of Montréal, from the ETH in Zurich, and from the AK Spiess for productive and fruitful collaborations. I also wish to acknowledge the excellent infrastructure of the MPI for Polymer Research, most notably Christian Bauer for invaluable technical support. Finally, I would like to say a big 'thank you' to my colleagues, friends, and family for making my life at the institute and beyond an enjoyable and unforgettable time.

Money is not important as long as you have it. I was in the lucky position to receive independent funding during my PhD thesis which enabled me to attend many conferences and workshops around the world. I gratefully acknowledge financial support from the Fonds of the German Chemical Industry through a Chemiefonds scholarship and from the Graduate School of Excellence "Materials Science in Mainz" funded by the German Research Foundation.

Santa Barbara, August 2011 Matthias J. N. Junk

Contents

Abbreviations and Acronyms

A	Hyperfine coupling tensor with principal elements A_{xx}, A_{yy}, and A_{zz} (orthorhombic symmetry) or $A_{\perp}$ and $A_{\parallel}$ (axial symmetry)
AIBN	2,2′-Azobis(isobutyronitrile)
a_{iso}	Isotropic hyperfine coupling constant
CAT1	TEMPO-4-trimethylammonium chloride
CW	Continuous wave
d	Dimensionality, thickness
δ	Chemical shift
Δ	Modulation depth
DEAAm	*N,N*-diethylacrylamide
DEER	Double electron–electron resonance
DI	Double integral
DNP	Dynamic nuclear polarization
DSA	Doxylstearic acid
DSC	Differential scanning calorimetry
DQ	Double quantum transition
EMAAm	*N,N*-ethylmethylacrylamide
EPR	Electron paramagnetic resonance
eq.	Equivalent, equation
ESE	Electron spin echo
ESEEM	Electron spin echo envelope modulation
EZ	Electron Zeeman
FA	Fatty acid(s)
FS	Force spectroscopy
FT	Fourier transform or Fourier transformation
FWHM	Full width at half maximum
g	**g** tensor with principal elements, g_{xx}, g_{yy}, and g_{zz} (orthorhombic symmetry) or $g_{\perp}$ and $g_{\parallel}$ (axial symmetry)
g_{iso}	Isotropic g-value
GPC	Gel permeation chromatography
HF	Hyperfine coupling

HSA	Human serum albumin
HYSCORE	Hyperfine sublevel correlation spectroscopy
I	Nuclear spin quantum number
k	Decay rate
λ	Inversion efficiency, spin–orbit coupling constant
LCST	Lower critical solution temperature
MAA	Methacrylic acid
MABP	Methacryloyloxybenzophenone
MD	Molecular dynamics
M_n, M_w	Number/weight average molecular weight
MRI	Magnetic resonance imaging
mw	Microwave
NiPAAm	*N*-isopropylacrylamide
NMR	Nuclear magnetic resonance
NQ	Nuclear quadrupole coupling
NZ	Nuclear Zeeman
PDB	Protein data bank
PFTFA	Pentafluorophenyltrifluoroacetate
PMMA	Polymethylmethacrylate
rDSA	Reduced doxylstearic acid (hydroxylamine)
rf	Radiofrequency
S	Electron spin quantum number, simulation
SA	Stearic acid
SL	Spin label
SNR	Signal-to-noise ratio
SQ	Single quantum transition
T	Absolute temperature (if not denoted otherwise)
T_1, T_2	Longitudinal/transverse spin relaxation time
τ_c	Rotational correlation time
T_C	Critical temperature
TEMPO	2,2,6,6-tetramethylpiperidine-1-oxyl
TEMPOL	4-hydroxy-TEMPO
T_g	Glass transition temperature
X-Band	Microwave frequency range of 8.5–10 GHz

Chapter 1
General Introduction

Amphiphilicity is a key structure forming element in many biological and synthetic systems. In the most general definition, it describes any chemical or structural contrast within a molecule, such as polar/non-polar, hydrocarbon/fluorocarbon, oligosiloxane/hydrocarbon or rigid/flexible. In this thesis, the amphiphilicity in its original definition is studied, namely the chimeric affinity of molecules for water due to hydrophilic and hydrophobic groups. In particular, amphiphiles have a tendency to self-assemble into larger structures due to partly non-favorable interactions with a solvent. This self-assembly results in their remarkable ability to encapsulate and transport small molecules. This so-called hosting is of central importance when a hydrophobic molecule needs to be transported through a polar medium or vice versa. The different variations of amphiphilic self-assembly and hosting constitute the central topic of this thesis.

Surfactants represent the prototype of amphiphilic systems. Due to their hydrophilic head group and hydrophobic tail, they self-assemble to vesicles, micelles, and a variety of other structures, where hydrophobic and hydrophilic subgroups segregate into distinct structural domains [1].

Bile acids are particularly interesting biological representatives of surfactants. Biosynthesized in the liver, they play a major role in the digestion of food as they emulsify fats and fat-soluble vitamins in the intestine. Exhibiting facial amphiphilicity, they include fats with their hydrophobic inner side, while the hydroxyl group-bearing side is exposed to the aqueous environment [2]. Biomimetic structural units that make use of these unique inclusion properties include chemically modified cholic acids [3, 4], cyclodextrines [5], and, in a more general sense, amphiphiles with biological functionalities [6–8].

Although low molecular-weight amphiphiles assume a variety of structures due to non-covalent interactions, it is their macromolecular analogues that exhibit the full power of amphiphilic self-assembly on length scales covering several orders of magnitude. In particular, the balance and interplay between hydrophilic and hydrophobic groups is optimized to perfection in biological macromolecules.

M. J. N. Junk, *Assessing the Functional Structure of Molecular Transporters by EPR Spectroscopy*, Springer Theses, DOI: 10.1007/978-3-642-25135-1_1,

In the case of DNA, both ionic and hydrophobic characteristics govern the self-assembly of nucleotides into the well-known double helical structure. Further, this helix can also be viewed as phase-segregated system with the base pairs located in its non-polar interior and hydrophilic phosphate units at the periphery [9]. As a second example, the specificity and catalytic activity of enzymes is triggered by structural subunits of distinct polarity, which are in this context called active sites and (hydrophobic) clefts [10–13].

In general, proteins are *the* natural paradigm for the influence of amphiphilicity on structure and function on a variety of length scales. Their primary hydrophobic and hydrophilic structure forming units, the amino acids [14, 15], self-assemble into α-helices or β-sheets, which often exhibit a hydrophobic and a hydrophilic face [16, 17]. Besides covalent linkages due to disulfide bonds, this facial amphiphilicity is the key driving force for their self-assembly to well-defined globular structures [18–20]. Hydrophobic interactions also account for the stabilization [21], organization [22], and repacking of the protein interior [23]. By changing the amphiphilicity, even a change of a secondary structure element was induced [24].

The complex structural motifs formed by amphiphilicity-driven self-assembly are crucial for the protein function [25, 26]. In particular, a whole class of proteins is designed and optimized for the transport of small molecules and ions. Transmembrane proteins like the well-known ion pump bacteriorhodopsin [27] and the sodium channel protein [28] span the phospholipid double layer of a cell and enable the exchange of various compounds. Other proteins are designed to actively transport small molecules in the blood circulatory system. Among the most prominent representatives, hemoglobin is responsible for the transport of oxygen and carbon dioxide [29], while albumin can complex a variety of endogenous ligands and drugs [30, 31].

It has been a long standing goal of chemists to mimic the unique features of biological macromolecules in synthetic polymers. Two different strategies were chosen to achieve an amphiphilic macromolecular structure. In the first approach, monomers with amphiphilic properties were polymerized or copolymerized to achieve desired structures like supramolecular assemblies [32, 33]. Among them, hydrophobic polyelectrolytes [34, 35] and acrylamides [36] have received considerable attention with *N*-isopropylacrylamide (NiPAAm) as the best studied representative [37]. Also, polymers with amphiphilic side chains have been reported [38, 39], which were proposed to act as drug carrier or as antimicrobial peptide mimics [40]. In a second approach, the self-assembly of block copolymers with hydrophilic and hydrophobic blocks [41] was successfully utilized to obtain vesicles [42, 43] and gels [44].

For the transport of small molecules, responsive amphiphilic polymers are of interest. In such systems, the structure can be switched from hydrophilic to hydrophobic upon the change of external parameters such as temperature [45] and pH [46, 47], or upon irradiation [32, 48], redox [49] or enzymatic [50, 51] reactions, to name only few possibilities [52]. Thus, amphiphilic, tunable polymeric gels and vesicles can be envisaged as targeted delivery agents, since they release water and incorporated molecules in the course of their dehydration [42, 53–56].

Finally, hyperbranched polymers and dendrimers are two closely related classes of macromolecules, which offer suitable features for the successful encapsulation of small molecules. While dendrimers were shown to entrap small molecules in their open core solely due to a restricted diffusion through their dense shell [57], nanocontainers for hydrophilic molecules were formed by amphiphilic hyperbranched core–shell polymers [58, 59].

The scope of this thesis is to elucidate how structure and dynamics govern the host and transport function of amphiphilic materials. Magnetic resonance techniques allow for a local, nanoscopic characterization of both structure and dynamics in disordered macromolecular systems [60, 61]. Specifically, electron paramagnetic resonance (EPR) spectroscopy offers structural information on a length scale ranging from atomic distances to ~8 nm and information on dynamic processes in the ps to μs regime [62–64].

Since EPR-active species in macromolecular systems are rare, structural information of the systems is achieved by adding small paramagnetic tracers, so-called spin probes, which are sensitive to the local viscosity and polarity [61, 65]. This method offers two distinct advantages. First, the interpretation of the data is largely facilitated, since only signals from the spin probes are detected. Second, the structure of the spin probes can be adapted to the problem of interest. It may be favorable to select a molecule that reflects the amphiphilicity of the studied systems or that can even act as tracer for a specific guest molecule when host–guest properties are studied. In either case, the structure of the system is characterized from the perspective of an incorporated guest molecule.

This thesis is divided into eight chapters. The fundamentals of the EPR spectroscopic methods applied are introduced in Chap. 2. In the subsequent five chapters, four different amphiphilic systems are characterized in terms of their functional structure, each system representing an important class of materials.

Human serum albumin (HSA) is a highly important carrier for drugs and endogenous compounds in the blood circulatory system [31, 66]. In Chap. 3, its highly relevant function as fatty acid transporter is studied. Though extensive crystallographic data on the binding of fatty acids to HSA exist [67–69], the approach presented here characterizes the protein structure from the ligand's point of view. In particular, this method can be used to assess, to which extent the crystal structure reflects the highly dynamic functional structure of the protein in solution.

In Chap. 4, the inclusion properties of triazole-modified star-shaped derivatives based on cholic acid are characterized. These biomimetic systems make use of the facial amphiphilicity of cholic acid to encapsulate and transport hydrophobic molecules in hydrophilic solvents and vice versa [70–72]. Here, Cu^{2+} is utilized as paramagnetic tracer to access structural information by its interaction with triazole moieties in the oligomer. Specifically, the influence of the relative position of the triazole in the oligomer on the binding affinity to heavy metal ions is elucidated.

In the following chapters, the focus shifts from biological to synthetic systems. The nanoscopic structure of a photocrosslinked hydrogel based on NiPAAm [73] during the thermal phase transition is determined in Chap. 5. Although omnipresent in polymer research, little effort has been applied to study how the changing solvent–solute interactions affect the thermoresponsive gel on the nanoscopic scale. In this study, an amphiphilic spin probe with affinities for both hydrophilic and hydrophobic polymer regions is used as a substitute for an incorporated host molecule. From this perspective, changes of the local structure are monitored with the focus on non-equilibrium phenomena such as dynamic and static heterogeneities.

Related to this, the use of thermoresponsive gels can facilitate method development in magnetic resonance. A derivative of this gel with covalently attached paramagnetic units (spin labels) is introduced as a hyperpolarization agent for nuclear magnetic resonance (NMR) spectroscopy (Chap. 6). With a method called dynamic nuclear polarization (DNP), the large polarization of electron spins is transferred to nuclear spins to achieve enhancement of the NMR signal [74]. It is proposed that the thermally triggered collapse enables a fast and efficient separation of the polarized solution from the toxic and relaxation-enhancing paramagnetic moieties.

In Chap. 7, the local structure of a more sophisticated thermoresponsive system is characterized. Dendronized polymers with oxyethylene building blocks exhibit a sharp and fully reversible phase transition in aqueous solution [75–77]. Above the critical temperature, they form well defined mesoglobules with a diameter of several hundred nanometers [78]. Applying the same amphiphilic spin probe as in Chap. 5, the local structural changes below and above the critical temperature are followed in terms of dehydration. The observed structural features on both nanoscopic and mesoscopic scales are then related to the chemical properties of the dendritic cores.

Finally, all results are summarized and concluded in Chap. 8. Special emphasis is placed on the structural and functional similarities and differences of the studied systems, which are united in their amphiphilic nature.

References

1. Zana R (ed) (2005) Dynamics of surfactant self-assemblies: micelles, microemulsions, vesicles and lyotropic phases, surfactant science series, vol 25. CRC Press, Boca Raton
2. Venkatesan P, Cheng Y, Kahne D (1994) J Am Chem Soc 116:6955–6956
3. Walker S, Sofia MJ, Kakarla R, Kogan NA, Wierichs L, Longley CB, Bruker K, Axelrod HR, Midha S, Babu S, Kahne D (1996) Proc Natl Acad Sci USA 93:1585–1590
4. Janout V, Lanier M, Regen SL (1996) J Am Chem Soc 118:1573–1574
5. Sallas F, Darcy R, (2008) Eur J Org Chem, 957–969
6. Lang S (2002) Curr Opin Colloid Interface Sci 7:12–20
7. Berti D (2006) Curr Opin Colloid Interface Sci 11:74–78
8. Lu JR, Zhao XB, Yaseen M (2007) Curr Opin Colloid Interface Sci 12:60–67
9. Gilbert HF (2000) Basic concepts in biochemistry: a student's survival guide. McGraw-Hill Book Co, NY

10. Schrag JD, Li YG, Cygler M, Lang DM, Burgdorf T, Hecht HJ, Schmid R, Schomburg D, Rydel TJ, Oliver JD, Strickland LC, Dunaway CM, Larson SB, Day J, McPherson A (1997) Structure 5:187–202
11. Chen Z-J, Pudas R, Sharma S, Smart OS, Juffer AH, Hiltunen JK, Wierenga RK, Haapalainen AM (2008) J Mol Biol 379:830–844
12. Ghosh D, Griswold J, Erman M, Pangborn W (2009) Nature 457:219–223
13. Gaiser OJ, Piotukh K, Ponnuswamy MN, Planas A, Borriss R, Heinemann U (2006) J Mol Biol 357:1211–1225
14. Nozaki Y, Tanford C (1971) J Biol Chem 246:2211–2217
15. Rose GD, Geselowitz AR, Lesser GJ, Lee RH, Zehfus MH (1985) Science 229:834–838
16. Perutz MF, Kendrew JC, Watson HC (1965) J Mol Biol 13:669–678
17. Eisenberg D, Weiss RM, Terwilliger TC (1982) Nature 299:371–374
18. Tanford C (1978) Science 200:1012–1018
19. Wolfenden R (1983) Science 222:1087–1093
20. Janin J (1979) Nature 277:491–492
21. Maynard AJ, Sharman GJ, Searle MS (1998) J Am Chem Soc 120:1996–2007
22. Rees DC, DeAntonio L, Eisenberg D (1989) Science 245:510–513
23. Sandberg WS, Terwilliger TC (1991) Proc Natl Acad Sci USA 88:1706–1710
24. Gordon DJ, Balbach JJ, Tycko R, Meredith SC (2004) Biophys J 86:428–434
25. Krone MG, Hua L, Soto P, Zhou RH, Berne BJ, Shea JE (2008) J Am Chem Soc 130:11066–11072
26. Roise D, Theiler F, Horvath SJ, Tomich JM, Richards JH, Allison DS, Schatz G (1988) EMBO J 7:649–653
27. Luecke H, Schobert B, Richter HT, Cartailler JP, Lanyi JK (1999) J Mol Biol 291:899–911
28. Catterall WA (2000) Neuron 26:13–25
29. Martin H, Nowicki L (eds) (1972) Synthese, Struktur und Funktion des Hämoglobins, Hämatologie und Bluttransfusion, Bd. 10. Lehmann, München
30. Peters T (1995) All about albumin: biochemistry genetics and medical applications. Academic Press, New York
31. Carter DC, Ho JX (1994) Adv Protein Chem 45:153–203
32. Ringsdorf H, Schlarb B, Venzmer J (1988) Angew Chem Int Ed 27:113–158
33. Kale TS, Klaikherd A, Popere B, Thayumanavan S (2009) Langmuir 25:9660–9670
34. Khokhlov AR, Khalatur PG (2005) Curr Opin Colloid Interface Sci 10:22–29
35. Kotz J, Kosmella S, Beitz T (2001) Prog Polym Sci 26:1199–1232
36. Cao Y, Zhao N, Wu K, Zhu XX (2009) Langmuir 25:1699–1704
37. Schild HG (1992) Prog Polym Sci 17:163–249
38. Liu K-H, Chen S-Y, Liu D-M, Liu T-Y (2008) Macromolecules 41:6511–6516
39. Brustolin F, Goldoni F, Meijer EW, Sommerdijk NAJM (2002) Macromolecules 35:1054–1059
40. Lienkamp K, Madkour AE, Musante A, Nelson CF, Nüsslein K, Tew GN (2008) J Am Chem Soc 130:9836–9843
41. Alexandridis P, Hatton TA (1995) Colloid Surf A 96:1–46
42. Discher DE, Eisenberg A (2002) Science 297:967–973
43. Gaucher G, Dufresne M-H, Sant VP, Kang N, Maysinger D, Leroux J-C (2005) J Controlled Release 109:169–188
44. Nowak AP, Breedveld V, Pakstis L, Ozbas B, Pine DJ, Pochan D, Deming TJ (2002) Nature 417:424–428
45. Yu H, Grainger DW (1993) J Appl Polym Sci 49:1553–1563
46. Gil ES, Hudson SA (2004) Prog Polym Sci 29:1173–1222
47. Haag R (2004) Angew Chem Int Ed 43:278–282
48. Haubs M, Ringsdorf H (1985) Angew Chem Int Ed 24:882–883
49. Saji T, Hoshino K, Aoyagui S (1985) J Am Chem Soc 107:6865–6868
50. Ghadiali JE, Stevens MM (2008) Adv Mater 20:4359–4363
51. Yang Z, Liang G, Xu B (2008) Acc Chem Res 41:315–343

52. Wang Y, Xu H, Zhang X (2009) Adv Mater 21:2849–2864
53. de las Heras Alarcón C, Pennadam S, Alexander C (2005) Chem Soc Rev 34:276–285
54. Qiu Y, Park K (2001) Adv Drug Del Rev 53:321–339
55. Chung JE, Yokoyama M, Yamato M, Aoyagi T, Sakurai Y, Okano T (1999) J Controlled Release 62:115–127
56. Soppimath KS, Liu LH, Seow WY, Liu SQ, Powell R, Chan P, Yang YY (2007) Adv Funct Mater 17:355–362
57. Jansen JFGA, de Brabander–van den Berg EMM, Meijer EW (1994) Science 266:1226–1229
58. Sunder A, Kramer M, Hanselmann R, Mülhaupt R, Frey H (1999) Angew Chem Int Ed 38:3552–3555
59. Chen Y, Shen Z, Pastor-Perez L, Frey H, Stiriba SE (2005) Macromolecules 38:227–229
60. Schmidt-Rohr K, Spiess HW (1994) Multidimensional solid-state NMR and polymers. Academic Press, London
61. Schlick S (ed) (2006) Advanced ESR methods in polymer research. Wiley-Interscience, Hoboken
62. Schweiger A, Jeschke G (2001) Principles of pulse electron paramagnetic resonance. Oxford University Press, Oxford
63. Jeschke G (2002) Macromol Rapid Commun 23:227–246
64. Hinderberger D, Jeschke G (2006) Site-specific characterization of structure and dynamics of complex materials by EPR spin probes. In: Webb GA (ed) Modern magnetic resonance. Springer, Berlin
65. Owenius R, Engstrom M, Lindgren M, Huber M (2001) J Phys Chem A 105:10967–10997
66. Curry S (2009) Drug Metabol Pharmacokinet 24:342–357
67. He XM, Carter DC (1992) Nature 358:209–215
68. Curry S, Mandelkow H, Brick P, Franks N (1998) Nat Struct Biol 5:827–835
69. Bhattacharya AA, Grüne T, Curry S (2000) J Mol Biol 303:721–732
70. Luo J, Chen Y, Zhu XX (2007) Synlett 14:2201–2204
71. Luo J, Chen Y, Zhu XX (2009) Langmuir 25:10913–10917
72. Janout V, Jing BW, Staina IV, Regen SL (2003) J Am Chem Soc 125:4436–4437
73. Beines PW, Klosterkamp I, Menges B, Jonas U, Knoll W (2007) Langmuir 23:2231–2238
74. Abragam A, Goldman M (1978) Rep Prog Phys 41:395–467
75. Li W, Zhang A, Feldman K, Walde P, Schlüter AD (2008) Macromolecules 41:3659–3667
76. Li W, Zhang A, Schlüter AD (2008) Chem Commun 5523–5525
77. Li W, Wu D, Schlüter AD, Zhang A (2009) J Polym. Sci Part A: Polym Chem 47:6630–6640
78. Bolisetty S, Schneider C, Polzer F, Ballauff M, Li W, Zhang A, Schlüter AD (2009) Macromolecules 42:7122–7128

Chapter 2
Electron Paramagnetic Resonance Theory

2.1 Historical Review

In 1921, Gerlach and Stern observed that a beam of silver atoms splits into two lines when it is subjected to a magnetic field [1–3]. While the line splitting in optical spectra, first found by Zeeman in 1896 [4, 5], could be explained by the angular momentum of the electrons, the *s*-electron of silver could not be subject to such a momentum, not to mention that an azimuthal quantum number $l = 1/2$ cannot be explained by classical physics. At that time, quantum mechanics was still an emerging field in physics and it took another three years until this anormal Zeeman effect was correctly interpreted by the joint research of Uhlenbeck, a classical physicist, and Goudsmit, a fellow of Paul Ehrenfest [6, 7]. They postulated a so-called 'spin', a quantized angular momentum, as an intrinsic property of the electron. This research marks the cornerstone of electron paramagnetic resonance (EPR) spectroscopy which is based on the transitions between quantized states of the resulting magnetic moment.

Cynically, the worst event in the twentieth century boosted the development of EPR spectroscopy as after World War II suitable microwave instrumentation was readily available from existing radar equipment. This lead to the observation of the first EPR spectrum by the Russian physicist Zavoisky in 1944 [8, 9] already one year before the first nuclear magnetic resonance (NMR) spectrum was recorded [10, 11]. The development of EPR and NMR went in the same pace during the first decade though NMR was by far more widely used. But in 1965, NMR spectroscopy experienced its final boost with the development of the much faster Fourier transform (FT) NMR technique which also opened the development of completely new methodologies in this field [12]. The corresponding pulse EPR spectroscopy suffered from expensive instrumentation, the lack of microwave components, sufficiently fast digital electronics and intrinsic problems of limited microwave power. Although the first pulse EPR experiment was reported by Blume already in 1958 [13], pulse EPR was conducted by only a small number of research groups over several decades.

M. J. N. Junk, *Assessing the Functional Structure of Molecular Transporters by EPR Spectroscopy*, Springer Theses, DOI: 10.1007/978-3-642-25135-1_2,

In the 1980s, the required equipment became cheaper and manageable and the first commercial pulse EPR spectrometer was released to the market [14], followed only ten years later by the first commercial high field spectrometer [15]. This development of equipment promoted the invention of new and the advancement of already existing methods. Nowadays, a vast EPR playground is accessible to an ever growing research community, which became as versatile as the spectroscopic technique itself.

2.2 EPR Fundamentals

2.2.1 Preface

While NMR is a standard spectroscopic technique for the structural determination of molecules, the closely related EPR spectroscopy is still sparsely known by many scientists. This discrepancy originates from the lack of naturally occurring paramagnetic systems due to the fact that the formation of a chemical bond is inherently coupled to the pairing of electrons and a resulting overall electron spin of $S = 0$.

This lack of EPR-active materials is both the biggest disadvantage and the biggest advantage of the method. On the one hand, the method is restricted to few existing paramagnetic systems, radicals and transition metal complexes with a residual electron spin. On the other hand, this selectivity turns out advantageous in the study of complex materials, since only few paramagnetic species lead to interpretable EPR spectra.

Since most materials are diamagnetic, paramagnetic species have to be artificially introduced as tracer molecules. Dependent on the manner of introduction, they are called spin labels or spin probes [16]. While spin labels are covalently attached to the system of interest, no chemical linkage is formed between spin probes and the material (see Sect. 2.6.1). In both cases, the tracer molecules are sensitive to their local surrounding in terms of structure and dynamics and thus allow an indirect *molecular* observation of the materials' properties [17, 18].

Additionally, dipole–dipole couplings between two electrons allow for distance measurements in the highly relevant range of 1.5–8 nm [19–21]. Attached to selected sites in biological and synthetic macromolecules, structural information of such complex materials becomes accessible [22–25]. This approach turns out especially powerful if the systems under investigation lack long-range order and scattering methods cannot be applied for structural analysis. Magnetic resonance techniques as intrinsically local methods do not require this constraint and EPR as one of only few methods allows for the structural characterization of these amorphous nanoscopic samples with a high selectivity and sensitivity [26–28].

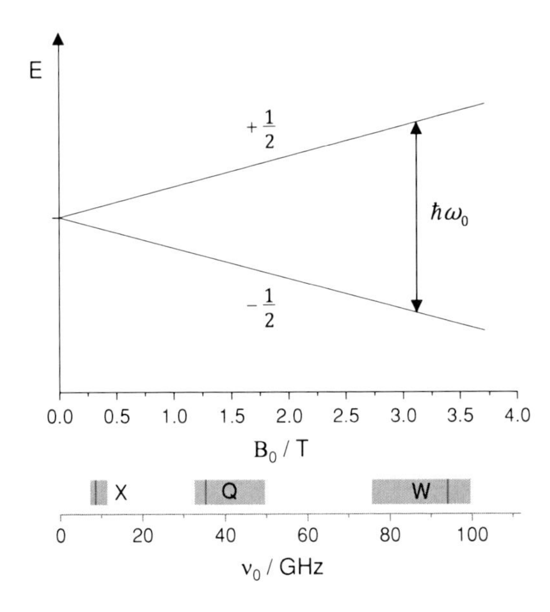

Fig. 2.1 Splitting of the energy levels of an electron spin subjected to a magnetic field with corresponding resonance frequencies, $g = g_e$, and typical EPR microwave bandsf

2.2.2 Resonance Phenomenon

Analogous to the orbital angular momentum **L**, the spin angular momentum **S** gives rise to a magnetic momentum **μ**

$$\boldsymbol{\mu} = h\gamma_e \mathbf{S} = -g\beta_e \mathbf{S} \tag{2.1}$$

with the magnetogyric ratio of the electron γ_e, the Bohr magneton $\beta_e = e\hbar/2m_e$, and the g-factor $g \approx 2$. The exact g-value for a free electron of $g_e = 2.0023193043617(15)$ is predicted by quantum electrodynamics and is the most accurately determined fundamental constant by both theory and experiment [29]. If the electron is subjected to an external magnetic field $\mathbf{B}_0^T = (0, 0, B_0)$,[1] the energy levels of the degenerate spin states split depending on their magnetic quantum number $m_S = \pm 1/2$ and the strength of the magnetic field B_0,

$$E = \pm \frac{1}{2} g\beta_e B_0. \tag{2.2}$$

Irradiation at a frequency ω_0, which matches the energetic difference ΔE between the two states, results in absorption,

$$\Delta E = \hbar\omega_0 = g\beta_e B_0. \tag{2.3}$$

[1] The correct term for **B** is magnetic induction. The term magnetic field originates from older magnetic resonance literature and is still commonly used while its symbol **H** was replaced by **B**.

The spectroscopic detection of this absorption is the fundamental principle of EPR spectroscopy. The resonance frequency $\omega_0 = -\gamma_e B_0$ is named Larmor frequency after J. Larmor, who described the analogous motion of a spinning magnet in a magnetic field in 1904. A schematic of the resonance phenomenon is depicted in Fig. 2.1.

For $g = 2$, magnetic fields between 0.1 and 1.5 Tesla, easily achievable with electro-magnets, result in resonance frequencies in the microwave (mw) regime between 2.8 and 42 GHz. Historically, mw frequencies are divided into bands. Most commercially available spectrometers operate at X-band ($\sim$9.5 GHz), Q-band ($\sim$36 GHz), or W-band ($\sim$95 GHz). All EPR measurements in this work were performed at X-band.

2.2.3 Magnetization

In general, EPR spectroscopy is conducted on a large ensemble of spins. The actual quantity detected is the net magnetic moment per unit volume, the macroscopic magnetization **M** (Eq. 2.7). The relative populations of the two energy states $|\alpha\rangle$ $(m_S = +1/2)$ and $|\beta\rangle$ $(m_S = -1/2)$ are given by the Boltzmann distribution

$$\frac{n_\alpha}{n_\beta} = \exp\left(-\frac{\Delta E}{kT}\right). \tag{2.4}$$

The excess polarization is described by the polarization P

$$P = \frac{n_\alpha - n_\beta}{n_\alpha + n_\beta} = \frac{1 - \exp(-\Delta E/kT)}{1 + \exp(-\Delta E/kT)}. \tag{2.5}$$

In a high temperature approximation ($\Delta E \ll kT$), which is valid for all experiments described in this thesis, the polarization is given by

$$P = \tanh\left(\frac{\hbar\gamma_e B_0}{2kT}\right) \approx \frac{\hbar\gamma_e B_0}{2kT}. \tag{2.6}$$

For a static magnetic field in z-direction, as described above, the equilibrium magnetization yields

$$\mathbf{M}_0 = \frac{1}{V}\sum_i \mu_i = \frac{1}{2}N\hbar\gamma_e P\mathbf{e}_z, \tag{2.7}$$

where $N = n_\alpha + n_\beta$ is the total number of spins. It is proportional to the magnetic field and inversely proportional to the temperature. Since the polarization is proportional to the magnetogyric ratio of the electron, the magnetization is proportional to γ_e^2, explaining the high sensitivity of EPR in comparison to NMR. Later in this thesis, the polarization difference between electrons and protons is utilized to enhance the magnitude of NMR signals (Chap. 6).

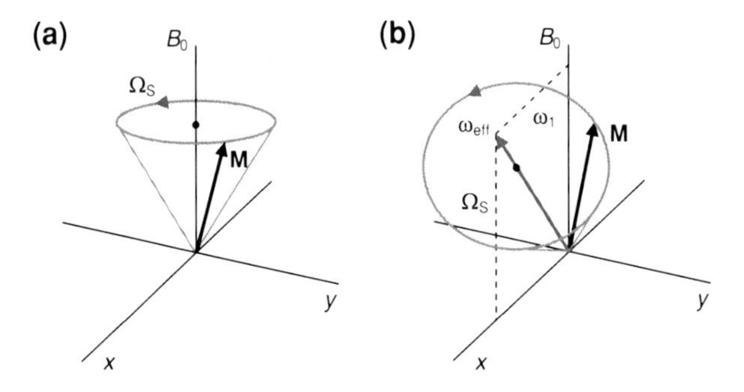

Fig. 2.2 **a** Free precession of the magnetization vector **M** in the rotation frame with the precession frequency $\Omega_S = \omega_0 - \omega_{mw}$. **b** Nutation of the magnetization vector during off-resonant mw irradiation with a circularly polarized mw field along x with amplitude ω_1. Reproduced from [76] by permission of Oxford University Press (www.uop.com)

2.2.4 Bloch Equations

Following the Larmor theorem, the motion of a magnetic moment in a magnetic field gives rise to a torque. For a single spin, this torque is given by

$$\hbar \frac{d\mathbf{S}}{dt} = \boldsymbol{\mu} \times \mathbf{B}. \tag{2.8}$$

Using the net magnetization **M** and dividing by V, we get

$$\hbar \frac{d\mathbf{M}}{dt} = \mathbf{M} \times \gamma_e \mathbf{B}. \tag{2.9}$$

When the magnetization is in equilibrium and a static magnetic field is applied along z, the magnetization vector is time-invariant and cannot be detected. However, any displacement from the z-axis results in a precessing motion around this axis with the Larmor frequency Ω_0, giving rise to a detectable alternating magnetic field. It is convenient to define a coordinate system that rotates counterclockwise with the microwave frequency Ω_{mw}. In this coordinate system, the magnetization vector rotates with a precession frequency $\Omega_S = \Omega_0 - \Omega_{mw}$ (Fig. 2.2a), which is the resonance offset between the Larmor and the mw frequency.

To detect an EPR signal, besides $\mathbf{B}_0$ an additional oscillating mw field $\mathbf{B}_1^T = (B_1 \cos(\omega_{mw} t), B_1 \sin(\omega_{mw} t), 0)$ is applied along x, which moves the magnetization vector away from its equilibrium position (Fig. 2.2b). For on-resonant mw irradiation ($\Omega_S = 0$), the effective nutation frequency ω_{eff} equals $\omega_1 = g\beta_e B_1/\hbar$ and the magnetization vector precesses around the x-axis while the magnetization vector is hardly affected if the mw frequency is far off-resonant ($\Omega_S \gg \omega_1$).

To fully describe the motion of the magnetization vector, relaxation effects also need to be considered. The longitudinal relaxation time T_1 characterizes the process(es) that make the magnetization vector return to its thermal equilibrium state while the transverse relaxation time T_2 describes the loss of coherence in the transverse plane due to spin–spin interactions. Felix Bloch first derived the famous equation of motion which fully describes the evolution of the magnetization [30]

$$\hbar \frac{d\mathbf{M}}{dt} = \mathbf{M}(t) \times \gamma_e \mathbf{B}(t) - \mathbf{R}(\mathbf{M}(t) - \mathbf{M}_0) = \begin{pmatrix} -\Omega_S M_y - M_x/T_2 \\ \Omega_S M_x - \omega_1 M_z - M_y/T_2 \\ \omega_1 M_y - (M_z - M_0)/T_1 \end{pmatrix}. \tag{2.10}$$

The magnetic field $\mathbf{B}$ consists of the static magnetic field $\mathbf{B}_0$ as well as the oscillating mw field $\mathbf{B}_1$,

$$\mathbf{B} = \begin{pmatrix} B_1 \cos(\omega_{\mathrm{mw}} t) \\ B_1 \sin(\omega_{\mathrm{mw}} t) \\ B_0 \end{pmatrix}, \tag{2.11}$$

and the relaxation tensor is given by

$$\mathbf{R} = \begin{pmatrix} T_2^{-1} & 0 & 0 \\ 0 & T_2^{-1} & 0 \\ 0 & 0 & T_1^{-1} \end{pmatrix}. \tag{2.12}$$

2.2.5 Continuous Microwave Irradiation

The most facile EPR experiment can be realized by continuous microwave irradiation of a sample placed in a magnetic field and detection of the microwave absorption. It is extremely difficult to produce a microwave source that provides a variable frequency range of several octaves with a sufficient amplitude and frequency stability. Hence, the microwave frequency (the quantum) is kept constant, while the magnetic field (the energy level separation) is varied.

A second EPR characteristic is determined by the instrumentation. The detector, a microwave diode, is sensitive to a broad frequency range. To reduce the frequency range of the detected noise, the EPR signal is modulated by a sinusoidal modulation of the magnetic field and only the modulated part of the diode output voltage is detected. Besides a drastically enhanced signal-to-noise ratio (SNR), this method implies the detection of the first derivative of the absorption spectrum rather than the absorption line itself. Although the detected signal is proportional to the modulation amplitude $\Delta B_0, \Delta B_0$ should not exceed one-third of the line width ΔB_{PP} (cf. Fig. 2.3) to avoid disturbed line shapes.

After a sufficiently long continuous microwave irradiation, the magnetization will reach a stationary state and the time derivatives of the magnetization vector vanish. In this case, the Bloch equation (Eq. 2.10) becomes a linear system of equations and the components of the magnetic field vector are given by

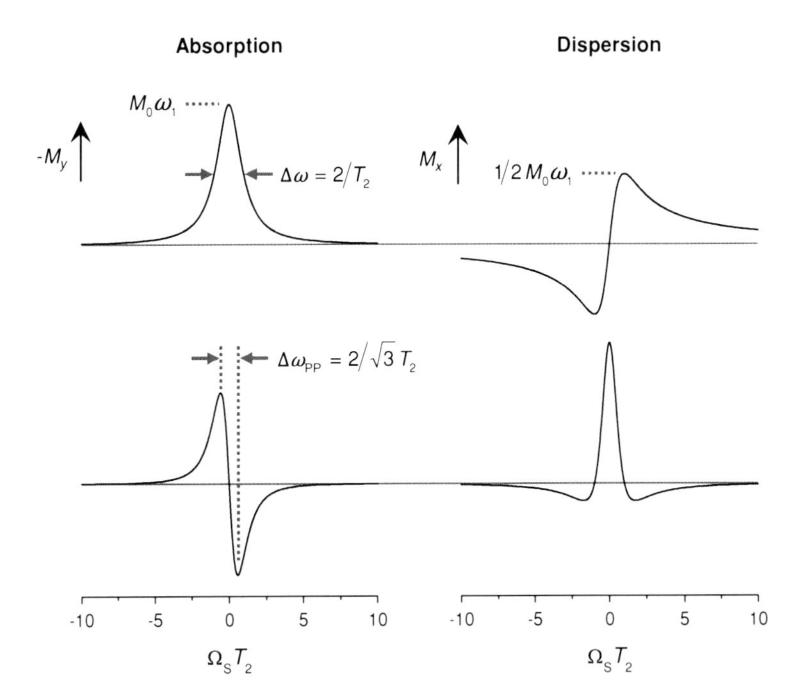

Fig. 2.3 *Top*: Lorentzian absorption and dispersion lineshapes calculated by Eqs. 2.15 and 2.16. *Bottom*: first derivative of the Lorentzian lines as observed in CW EPR spectroscopy

$$M_x = M_0\omega_1 \frac{\Omega_S T_2^2}{1 + \Omega_S^2 T_2^2 + \omega_1^2 T_1 T_2}, \tag{2.13a}$$

$$M_y = -M_0\omega_1 \frac{T_2}{1 + \Omega_S^2 T_2^2 + \omega_1^2 T_1 T_2}, \tag{2.13b}$$

$$M_z = M_0 \frac{1 + \Omega_S^2 T_2^2}{1 + \Omega_S^2 T_2^2 + \omega_1^2 T_1 T_2}. \tag{2.13c}$$

M_Z is called longitudinal magnetization and cannot be detected with conventional experimental setups. The components M_x and M_y are called transverse magnetization or coherence and can be measured simultaneously in a quadrature-detection scheme with two microwave reference signals phase shifted by 90° with respect to each other. In this case, a complex signal

$$V = -M_y + iM_x \tag{2.14}$$

is obtained.

For low microwave powers, i.e. $\omega_1^2 T_1 T_2 \ll 1$, the last term of the denominator in Eqs. 2.11–2.13 vanishes. In this *linear regime*, M_x and M_y are proportional to ω_1. The real part of the complex signal amounts to

$$-M_y = M_0\omega_1 \frac{T_2}{1+\Omega_S^2 T_2^2} = M_0\omega_1 \frac{T_2^{-1}}{T_2^{-2}+\Omega_S^2}. \tag{2.15}$$

This corresponds to a Lorentzian absorption line with a width T_2^{-1} and an amplitude $M_0\omega_1$. The imaginary part is given by

$$M_x = M_0\omega_1 \frac{\Omega_S T_2^2}{1+\Omega_S^2 T_2^2} = M_0\omega_1 \frac{\Omega_S}{T_2^{-2}+\Omega_S^2}, \tag{2.16}$$

corresponding to a Lorentzian dispersion line. Both line shape functions as well as their first derivatives are illustrated in Fig. 2.3. Since the dispersion line suffers from broad flanks and decreased amplitudes, only absorption lines are recorded, which offer a better SNR and a better resolution in presence of multiple lines.

In the analysis of first derivative spectra, the definition of a peak-to-peak line width Γ_{PP} is favorable. It is related to the full width at half maximum (FWHM) of the respective absorption line by

$$\frac{\Gamma_{PP}}{\Gamma_{FWHM}} = 3^{-1/2}. \tag{2.17}$$

On an angular frequency scale, the peak-to-peak line width is given by $\Delta\omega_{PP} = 2/\sqrt{3}T_2$. In magnetic field swept spectra, the relation

$$\Delta B_{PP} = \frac{2}{\sqrt{3}T_2}\frac{\hbar}{g\beta_e} \tag{2.18}$$

holds when the spectrum is not inhomogeneously broadened by unresolved hyperfine couplings. In the latter case, the spectrum consists of Gauss or Voigt lines rather than of pure Lorentzian signals.

EPR is a spectroscopic method relying on a purely quantum mechanical construct, the electron spin. So far, this has been described in the picture of classical physics. In the next section, we will proceed to a quantum-mechanical description, and to the spin Hamiltonian which describes all magnetic interactions of the spin with its environment.

2.3 Types of Interactions and Spin Hamiltonian

The spin Hamiltonian can be derived by the Hamiltonian of the whole system by separating the energetic contributions involving the spin from all other contributions [31]. It contains all interactions of the electron spin with the external magnetic field and internal magnetic moments i.e. other spins in the vicinity of the electron spin. Dependent on the number of interacting spins J, the spin Hamiltonian spans a Hilbert space with the dimension

$$n_H = \prod_k (2J_k + 1), \tag{2.19}$$

which describes the number of energy levels of the system. The energy of a paramagnetic species in the ground state with an effective electron spin S and n coupled nuclei with spins I is described by the static spin Hamiltonian

$$\mathcal{H}_0 = \mathcal{H}_{\mathrm{EZ}} + \mathcal{H}_{\mathrm{ZFS}} + \mathcal{H}_{\mathrm{HF}} + \mathcal{H}_{\mathrm{NZ}} + \mathcal{H}_{\mathrm{NQ}} + \mathcal{H}_{\mathrm{NN}}. \tag{2.20}$$

The terms describe the electron Zeeman interaction $\mathcal{H}_{\mathrm{EZ}}$, the zero-field splitting $\mathcal{H}_{\mathrm{ZFS}}$, hyperfine couplings between the electron spin and the nuclear spins $\mathcal{H}_{\mathrm{HF}}$, the nuclear Zeeman interactions $\mathcal{H}_{\mathrm{NZ}}$, the quadrupolar interactions $\mathcal{H}_{\mathrm{NQ}}$ for nuclear spins with $I > 1/2$, and spin–spin interactions between pairs of nuclear spins $\mathcal{H}_{\mathrm{NN}}$. The different terms of the spin Hamiltonian in Eq. 2.20 are ordered according to their typical energetic contribution. All energies will be given in angular frequency units.

2.3.1 Electron Zeeman Interaction

The interaction between the electron spin and the external magnetic field is described by the electron Zeeman term

$$\mathcal{H}_{\mathrm{EZ}} = \frac{\beta_{\mathrm{e}}}{\hbar}\mathbf{B}_0^{\mathrm{T}}\mathbf{g}\mathbf{S}, \tag{2.21}$$

which is the dominant term of the spin Hamiltonian for usually applied magnetic fields (high field approximation). Since both spin operator $\mathbf{S}$ and the external magnetic field $\mathbf{B}_0$ are explicitly orientation-dependent, $\mathbf{g}$ assumes the general form of a tensor with the components

$$\mathbf{g} = \begin{pmatrix} g_{xx} & g_{xy} & g_{xz} \\ g_{yx} & g_{yy} & g_{yz} \\ g_{zx} & g_{zy} & g_{zz} \end{pmatrix}. \tag{2.22}$$

It can be diagonalized via Euler angle transformation of the magnetic field vector into the molecular coordinate system of the radical to yield

$$\mathbf{g} = \begin{pmatrix} g_{xx} & 0 & 0 \\ 0 & g_{yy} & 0 \\ 0 & 0 & g_{zz} \end{pmatrix}. \tag{2.23}$$

The deviation of the $\mathbf{g}$ principal values from the g_{e} value of the free electron spin and its orientation dependence is caused by the spin–orbit coupling. Since the orbital angular momentum $\mathbf{L}$ is quenched for a non-degenerate ground state, only the interaction of excited states and ground state leads to an admixture of the orbital angular momentum to the spin angular momentum. The $\mathbf{g}$ tensor can be expressed by [32]

$$\mathbf{g} = g_{\mathrm{e}}\mathbf{1} + 2\lambda\mathbf{\Lambda} \tag{2.24}$$

with the spin–orbit coupling constant λ and the symmetric tensor $\mathbf{\Lambda}$ with elements

$$\Lambda_{ij} = \sum_{n \neq 0} \frac{\langle \psi_0 | L_i | \psi_n \rangle \langle \psi_n | L_j | \psi_0 \rangle}{\epsilon_0 - \epsilon_n}. \tag{2.25}$$

Each element Λ_{ij} describes the interactions of the SOMO ground state ψ_0 with energy ε_0 and the nth excited state ψ_n with energy ε_n. A large deviation from g_e results from a small energy difference between the SOMO and the lowest excited state and a large the spin–orbit coupling. For most organic radicals, the excited states are high in energy and $g_{jj} \approx g_\mathrm{e}$. Larger deviations are observed for transition metal complexes, which also benefit from the fact that the spin–orbit coupling as a relativistic effect is proportional to the molecular mass of the atom.

In solution, the orientation dependence of the **g** tensor is averaged by fast molecular motion and an isotropic g-value is observed, which amounts to

$$g_\mathrm{iso} = \frac{1}{3}\left(g_{xx} + g_{yy} + g_{zz}\right). \tag{2.26}$$

2.3.2 *Nuclear Zeeman Interaction*

Analogous to the electron Zeeman interaction, nuclear spins couple to the external magnetic field. This contribution is described by the nuclear Zeeman term

$$\mathcal{H}_\mathrm{NZ} = -\frac{\beta_\mathrm{n}}{\hbar} \sum_k g_{\mathrm{n},k} \mathbf{B}_0^\mathrm{T} \mathbf{I}_k. \tag{2.27}$$

The spin quantum number I and the g_n factor are inherent properties of a nucleus. In most experiments, the nuclear Zeeman interaction can be considered isotropic. It hardly influences EPR spectra, however affects nuclear frequency spectra measured by EPR techniques (cf. Sect. 2.10).

2.3.3 *Hyperfine Interaction*

The hyperfine interaction is one of the most important sources of information in EPR spectroscopy. It characterizes interactions between the electron spin and nuclear spins in its vicinity. Hence, it provides information about the direct magnetic environment of the spin. Its contribution to the Hamiltonian is given by

$$\mathcal{H}_\mathrm{HF} = \sum_k \mathbf{S}^\mathrm{T} \mathbf{A}_k \mathbf{I}_k = \mathcal{H}_\mathrm{F} + \mathcal{H}_\mathrm{DD}. \tag{2.28}$$

$\mathbf{A}$ is the hyperfine coupling tensor and $\mathbf{I}_k$ the spin operator of the kth coupled nucleus. This Hamiltonian can be further subdivided in an isotropic part $\mathcal{H}_\mathrm{F}$ and a dipolar part $\mathcal{H}_\mathrm{DD}$.

The nuclear magnetic moment gives rise to a dipole–dipole interaction between electron and nuclear spin, which acts through space. In general, the interaction between two magnetic dipoles $\boldsymbol{\mu}_1$ and $\boldsymbol{\mu}_2$ is given by

$$E = \frac{\mu_0}{4\pi}\frac{1}{r^3}\left[\boldsymbol{\mu}_1^\mathrm{T}\boldsymbol{\mu}_2 - \frac{3}{r^2}\left(\boldsymbol{\mu}_1^\mathrm{T}\mathbf{r}\right)\left(\boldsymbol{\mu}_2^\mathrm{T}\mathbf{r}\right)\right]. \tag{2.29}$$

With the introduction of the dipolar coupling tensor $\mathbf{T}$, the dipolar term of the hyperfine interaction can be expressed by

$$\mathcal{H}_\mathrm{DD} = \sum_k \mathbf{S}^\mathrm{T}\mathbf{T}_k\mathbf{I}_k. \tag{2.30}$$

In the hyperfine principal axis system, $\mathbf{T}$ is approximately described by

$$\mathbf{T} = \frac{\mu_0}{4\pi}\frac{g_\mathrm{e}g_\mathrm{n}\beta_\mathrm{e}\beta_\mathrm{n}}{R^3}\begin{pmatrix} -1 & & \\ & -1 & \\ & & 2 \end{pmatrix} = \begin{pmatrix} -T & & \\ & -T & \\ & & 2T \end{pmatrix}. \tag{2.31}$$

This representation neglects $\mathbf{g}$ anisotropies and spin–orbit couplings but is a good approximation as long as both effects are small. Since $\mathbf{T}$ is traceless, the dipolar part of the hyperfine interaction is averaged to zero by fast and isotropic rotation of the radical. In this case, only the isotropic part of the hyperfine interaction prevails. The energetic contributions of this so-called Fermi contact term

$$\mathcal{H}_\mathrm{F} = \sum_k a_{\mathrm{iso},k}\mathbf{S}^\mathrm{T}\mathbf{I}_k \tag{2.32}$$

are characterized by the isotropic hyperfine coupling constant

$$a_\mathrm{iso} = \frac{2}{3}\frac{\mu_0}{\hbar}g_\mathrm{e}g_\mathrm{n}\beta_\mathrm{e}\beta_\mathrm{n}|\psi_0(0)|^2. \tag{2.33}$$

This contribution originates, since an electron in a s-orbital possesses a finite electron spin density at the nucleus, $|\psi_0(0)|^2$. Via configuration interaction or spin polarization mechanisms, also electrons in orbitals with $l \neq 0$ contribute to the spin density at the nucleus and to the isotropic hyperfine coupling term [33].

2.3.4 *Nuclear Quadrupole Interaction*

Nuclei with $I \geq 1$ are characterized by a non-spherical charge distribution, which gives rise to a nuclear electrical quadrupole moment Q. This moment interacts with the electric field gradient at the nucleus, caused by electrons and nuclei in its vicinity. With the traceless nuclear quadrupole tensor **P**, this contribution is described by

$$\mathcal{H}_{\mathrm{NQ}} = \sum_{I_k > 1/2} \mathbf{I}_k^{\mathrm{T}} \mathbf{P}_k \mathbf{I}_k. \tag{2.34}$$

In EPR spectra, nuclear quadrupole interactions cause resonance shifts and the appearance of forbidden transitions. These small second-order effects are difficult to observe. In nuclear frequency spectra recorded by EPR, quadrupole couplings manifest themselves as first-order splittings (cf. Sect. 4.3.2).

2.3.5 Nuclear Spin–Spin Interaction

The dipole–dipole interaction between two nuclear spins is given by

$$\mathcal{H}_{\mathrm{NQ}} = \sum_{I_k > 1/2} \mathbf{I}_i^{\mathrm{T}} \mathbf{d}^{(i,k)} \mathbf{I}_k. \tag{2.35}$$

The nuclear dipole coupling tensor $\mathbf{d}^{(i,k)}$ provides the main information source in solid-state NMR [26]. However, the coupling is far too weak to be observed in EPR spectra and usually it is not even resolved in nuclear frequency spectra.

2.3.6 Zero-Field Splitting

For spin systems with a group spin $S > 1/2$ and non-cubic symmetry, the dipole–dipole couplings between individual electron spins remove the energetic degeneracy of the ground state. This zero-field splitting causes an energy splitting even in the absence of an external magnetic field and is expressed by

$$\mathcal{H}_{\mathrm{ZFS}} = \mathbf{S}^{\mathrm{T}} \mathbf{D} \mathbf{S} \tag{2.36}$$

with the symmetric and traceless zero-field interaction tensor **D** and the group spin $\mathbf{S} = \sum_k \mathbf{S}_k$. In this thesis, only paramagnetic species with $S = 1/2$ are studied in detail. Thus, the zero-field splitting term of the spin Hamiltonian can be neglected.

2.3.7 Weak Coupling Between Electron Spins

Strongly interacting electron spins are characterized by a group spin, as discussed in the previous section. A system of two weakly coupled unpaired electrons is more conveniently described by two single spin Hamiltonians and additional terms, which arise from the coupling,

$$\mathcal{H}_0(S_1, S_2) = \mathcal{H}_0(S_1) + \mathcal{H}_0(S_2) + \mathcal{H}_{\text{exch}} + \mathcal{H}_{\text{dd}}. \quad (2.37)$$

These excess terms characterize the contributions due to spin exchange $\mathcal{H}_{\textbf{exch}}$ and dipole–dipole coupling $\mathcal{H}_{\text{dd}}$ [34].

2.3.7.1 Heisenberg Exchange Coupling

When two species approach each other close enough, the orbitals of the two spins overlap and the unpaired electrons can be exchanged. In solids, the upper limit of 1.5 nm can be exceeded in strongly delocalized systems [22]. In liquid solutions, the exchange interaction is mainly governed by the collision of paramagnetic species leading to strongly overlapping orbitals for short times.

Exchange coupling can be differentiated into an isotropic and anisotropic contribution and is characterized by the exchange coupling tensor $\mathbf{J}$. Its contribution to the static spin Hamiltonian is given by

$$\mathcal{H}_{\text{exch}} = \mathbf{S}_1^{\mathrm{T}} \mathbf{J} \mathbf{S}_2 = J_{12} \mathbf{S}_1^{\mathrm{T}} \mathbf{S}_2 \quad (2.38)$$

The last term of Eq. 2.38 only applies for organic radicals, since the anisotropic part of the exchange tensor $\mathbf{J}$ can be neglected. In a system $S_1 = S_2 = 1/2$, the isotropic part can be described in terms of chemical bonding. Ferromagnetically coupled spins form a weakly bonded triplet state with $S = 1$, while a weakly antibonded singulet state ($S = 0$) is formed in case of antiferromagnetic coupling.

2.3.7.2 Dipole–Dipole Interaction

The dipole–dipole interaction between two electrons is treated in analogy to the dipolar coupling between an electron and a nucleus (Sect. 2.3.3). Since the measurement of dipolar couplings is a powerful tool to extract distance information and will be utilized extensively in the next section, we shall have a closer look on the inner working of the equations.

The interaction of two classic dipoles was already given in Eq. 2.29. Distributed randomly in space, as depicted in Fig. 2.4 (left), the interaction energy depends on their relative orientation to the connecting vector and can be calculated by

$$E = \frac{\mu_0}{4\pi} \frac{1}{r_{12}^3} (2 \cos\theta_1 \cos\theta_2 - \sin\theta_1 \sin\theta_2 \cos\chi). \quad (2.39)$$

In the high field approximation, the dipolar coupling to the external magnetic field dominates all other contributions. Hence, the dipoles align parallel to $\mathbf{B}_0$ (Fig. 2.4 right), and Eq. 2.39 simplifies to

$$E = \frac{\mu_0}{4\pi} \frac{1}{r_{12}^3} \left(1 - 3\cos^2\theta\right). \quad (2.40)$$

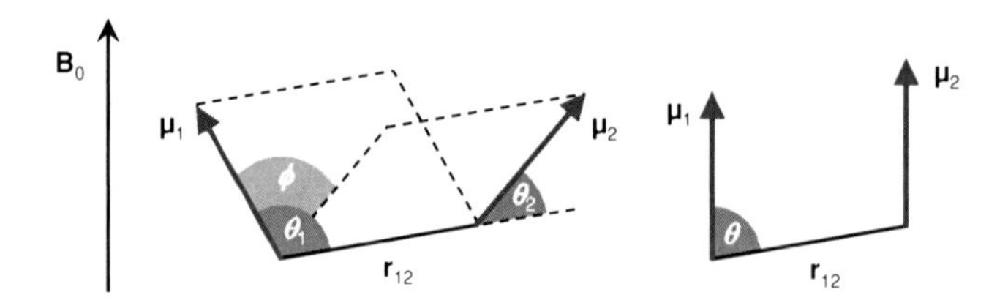

Fig. 2.4 Schematic representation of two interacting magnetic dipoles. *Left*: arbitrary orientation of the dipoles in space. The angles between the dipoles and the connection vector **r** are denoted θ_1 and θ_2. The angular offset between the two spanned planes is characterized by the dihedral angle χ. *Right*: orientation of both dipoles parallel to the external magnetic field $\mathbf{B}_0$. Adapted from [79] with permission from the author

In this approximation and neglecting **g** anisotropy, the spin Hamiltonian as a quantum-mechanical analogue amounts to

$$\mathcal{H}_{\mathrm{dd}} = \mathbf{S}^{\mathrm{T}}\mathbf{D}\mathbf{S} = \frac{\mu_0}{4\pi\hbar}\frac{1}{r_{12}^3} g_1 g_2 \beta_e^2 \left[\mathbf{S}_1^{\mathrm{T}}\mathbf{S}_2 - \frac{3}{r_{12}^2}\left(\mathbf{S}_1^{\mathrm{T}}\mathbf{r}_{12}\right)\left(\mathbf{S}_2^{\mathrm{T}}\mathbf{r}_{12}\right)\right]. \tag{2.41}$$

The dipolar coupling tensor **D** can be expressed in its principal axes frame as

$$\mathbf{D} = \frac{\mu_0}{4\pi\hbar}\frac{g_1 g_2 \beta_e^2}{r_{12}^3}\begin{pmatrix} -1 & & \\ & -1 & \\ & & 2 \end{pmatrix} = \begin{pmatrix} -\omega_{\mathrm{DD}} & & \\ & -\omega_{\mathrm{DD}} & \\ & & 2\omega_{\mathrm{DD}} \end{pmatrix}. \tag{2.42}$$

Since the frequency is proportional to r_{12}^{-3}, the distance between the coupled spin pair can be retrieved. This is discussed in detail in Sect. 2.11, where the corresponding pulse EPR method is introduced.

2.4 Anisotropy in EPR Spectra

Many interactions in magnetic resonance are anisotropic or have anisotropic components. In disordered samples, i.e. non-crystalline materials lacking long-range order, the radicals are distributed randomly with respect to the external magnetic field. If the rotational motion does not average the orientation of the paramagnetic species, these anisotropic interactions lead to powder spectra. For a given microwave frequency, spins fulfill the resonance condition at different magnetic field positions depending of their orientation. This leads to broadened spectra with characteristic line shapes.

2.4.1 g Anisotropy

As stated in Sect. 2.3, the Electron-Zeeman interaction depends on the absolute orientation of the molecule with respect to the external magnetic field.

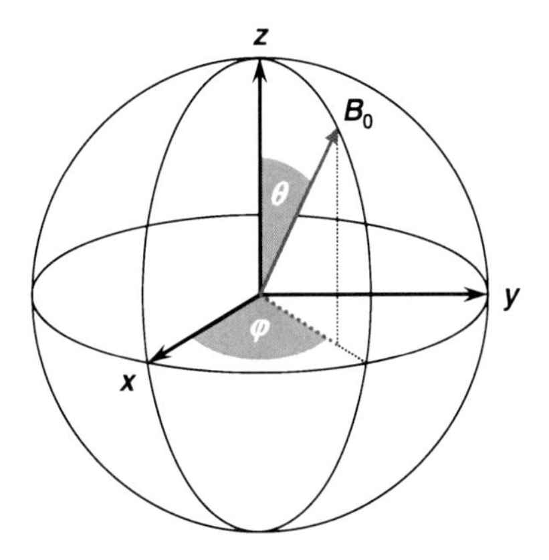

Fig. 2.5 Definition of the elevation angle θ and the azimuth φ in a unit sphere depicting the molecular coordinate system (collinear to the principal elements of the diagonalized **g** tensor, g_{xx}, g_{yy}, and g_{zz}) and the magnetic field vector $\mathbf{B}_0$

The orientation of the molecule can be characterized by the elevation angle θ and the azimuth φ of a spherical coordinate system. θ characterizes the angle between the molecular z-axis and the vector of the magnetic field $\mathbf{B}_0$. φ denotes the angle between its projection on the molecular xy-plane and the x-axis (Fig. 2.5).

Each point of the anisotropic spectrum is characterized by an effective g-value by the relation

$$B_0 = \frac{\hbar\omega_0}{g_{\text{eff}}\beta_{\text{e}}}. \tag{2.43}$$

In general, the effective g-value depends on both spherical angles and can be calculated by

$$g(\theta, \varphi) = (g_{xx}^2 \sin^2\theta \cos^2\varphi + g_{yy}^2 \sin^2\theta \sin^2\varphi + g_{zz}^2 \cos^2\theta)^{1/2}. \tag{2.44}$$

In case of axial symmetry with $g_{zz} = g_{\parallel}$ aligned parallel to the unique axis of the molecular frame and $g_{xx} = g_{yy} = g_{\perp}$ aligned perpendicular to this axis, the **g** tensor is given by

$$\mathbf{g} = \begin{pmatrix} g_{\perp} & 0 & 0 \\ 0 & g_{\perp} & 0 \\ 0 & 0 & g_{\parallel} \end{pmatrix}. \tag{2.45}$$

Hence, the φ dependence vanishes and the effective g-value amounts to

$$g(\theta) = (g_{\perp}^2 \sin^2\theta + g_{\parallel}^2 \cos^2\theta)^{1/2}. \tag{2.46}$$

Characteristic anisotropically broadened powder spectra for paramagnetic species with axial and orthorhombic symmetries are displayed in Fig. 2.6. Unit spheres illustrate the orientational contributions at the principal g-values. Since the intensity scales with the number of contributing orientations, the maximum spectral intensity is observed at effective g-values corresponding to $g_{\perp}$ and g_{yy}, respectively.

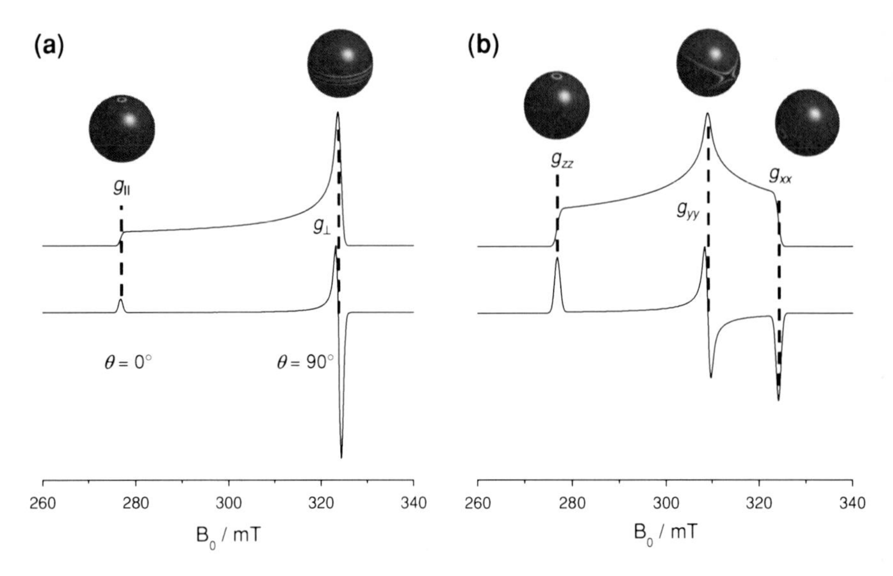

Fig. 2.6 Simulated absorption (*top*) and first derivative (*bottom*) powder spectra with anisotropic **g** tensors at a microwave frequency of 9.3 GHz. **a** Axial symmetry with $g_{\perp} = 2.05$ and $g_{\|} = 2.4$. **b** Orthorhombic symmetry with $g_{xx} = 2.05$, $g_{yy} = 2.15$, and $g_{zz} = 2.4$. Red spots on the unit spheres mark orientations that contribute to the spectra at magnetic field positions corresponding to the principal g-values upon excitation by a 32 ns mw pulse. Adapted from [79] with permission from the author

2.4.2 Combined Anisotropies in Real Spectra

Besides the g-value, both the hyperfine coupling constant and the zero–field splitting constant have anisotropic components which influence the line shape in CW EPR spectra. The explicit orientation dependence of dipole–dipole interactions between different unpaired electrons and its consequence for the DEER experiment is treated in Sects. 2.11 and 3.4.

In this thesis, no paramagnetic species with an effective electron spin $S > 1/2$ is studied which would give rise to zero-field splitting. However, the hyperfine anisotropy plays an important role for many radicals. It is given by the magnitude of the dipole–dipole interaction between the electron and the nuclear spin.

Simulated powder spectra of a Jahn–Teller distorted copper(II) complex and a nitroxide are shown in Fig. 2.7, which exhibit both **g** and **A** anisotropy. The **g** tensor of the copper complex is axially symmetric due to four identical equatorial ligands, while the nitroxide shows unique orientational dependences along all molecular axes. The hyperfine anisotropy, however, is approximately axially symmetric in both cases. In either case, the principal axes of the **g** and the **A** tensor coincide. A simplified explanation for the observed hyperfine anisotropy can be given as follows: The majority of the electron density is distributed in molecular orbitals aligned parallel to the molecular z-axis of the paramagnetic moiety (see also Sect. 2.6.2). Thus, the dipolar coupling with the proximate nuclei (^{63}Cu with $I = 3/2$ and ^{14}N with $I = 1$, respectively) in this direction dominates and gives rise to the highest hyperfine coupling, $A_{\parallel}$ or A_{zz}. The significantly lower hyperfine coupling values along the x- and the y-direction are not fully resolved.

The superposition of two spectral anisotropies accounts for the decreased orientational selectivity in comparison to the purely **g** anisotropic spectra displayed in Fig. 2.6.

2.5 Dynamic Exchange

In this section, we will take a closer look on how the spectral parameters are affected when a paramagnetic species undergoes a transition on the EPR timescale ($\sim$ns) that changes its electronic structure. This transition can be either a chemical reaction or any process that changes the conformation of the radical or its environment (see Sect. 2.6.4). In all cases, the resonance frequency of the EPR transition is affected due to a change of the hyperfine coupling, or the **g** tensor, or both. Additionally, the line width is affected by the average time frame during which this process happens.

This process will be referred to as *dynamic exchange* throughout this thesis since it is the dynamic transition of a spin probe between two different environments that is responsible for the spectral effects observed in Chap. 7. In magnetic resonance textbooks, the process is commonly called *chemical exchange* and treated analogously.

If we consider a transition from a species A to a species B with the rate constants forward and backward, k_1 and k_{-1},

$$\mathrm{A} \underset{k_{-1}}{\overset{k_{-1}}{\rightleftharpoons}} \mathrm{B}, \tag{2.47}$$

the average lifetimes of both species are given by

$$\tau_\mathrm{A} = \frac{1}{k_1} \quad \text{and} \quad \tau_\mathrm{B} = \frac{1}{k_{-1}}. \tag{2.48}$$

At equilibrium, the rates for the transition back and forward must be equal, so that the fractions ρ_j of each site are related to the average lifetimes by

$$\frac{\rho_\mathrm{A}}{\tau_\mathrm{A}} = \frac{\rho_\mathrm{B}}{\tau_\mathrm{B}}. \tag{2.49}$$

For a two-site problem, the equilibrium constant amounts to

$$K = \frac{\tau_\mathrm{B}}{\tau_\mathrm{A}} = \frac{\rho_\mathrm{B}}{\rho_\mathrm{A}} \tag{2.50}$$

and the fractions ρ_j can be expressed in terms of the average lifetimes by

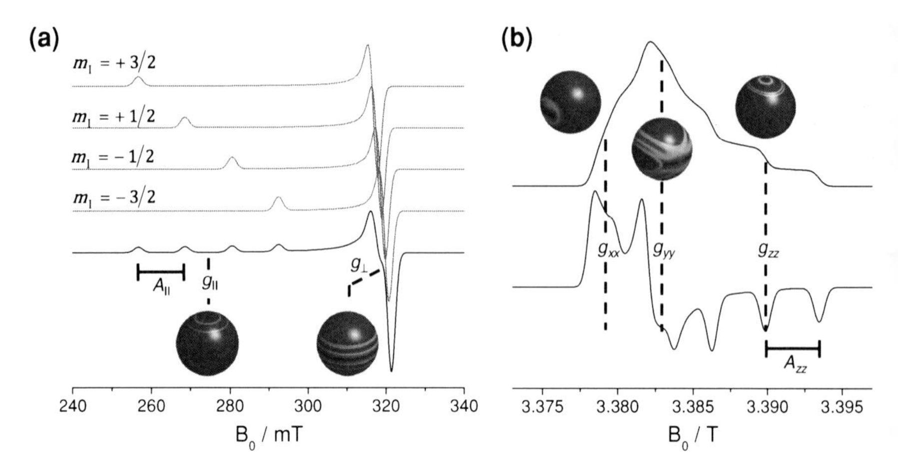

Fig. 2.7 Simulated EPR spectra for paramagnetic species with several anisotropic contributions. **a** Copper(II) spectrum with axial symmetric and collinear **g** and **A** tensors ($g_\perp = 2.085$, $g_{\|} = 2.42$ and $A_\perp = 28.0$ MHz, $A_{\|} = 403.6$ MHz) at a microwave frequency of 9.3 GHz. A 100% isotopic abundance of ^{63}Cu is assumed. Hypothetical spectra corresponding to different quantum numbers of the copper nucleus are displayed in gray. **b** W-band (95 MHz) absorption (*top*) and first derivative (*bottom*) EPR spectrum of a nitroxide radical with an orthorhombic **g** tensor ($g_{xx} = 2.0087$, $g_{yy} = 2.0065$, and $g_{zz} = 2.0023$) and a collinear hyperfine coupling tensor $A_{xx} = A_{yy} = 18.2$ MHz and $A_{zz} = (100.9$ MHz). Excited orientations are illustrated in red on unit spheres at principal g-value field positions. Note the decreased orientational selectivity in comparison to Fig. 2.6 due to the additional hyperfine contribution

$$\rho_\mathrm{A} = \frac{\tau_\mathrm{A}}{\tau_\mathrm{A} + \tau_\mathrm{B}} \quad \text{and} \quad \rho_\mathrm{B} = \frac{\tau_\mathrm{B}}{\tau_\mathrm{A} + \tau_\mathrm{B}}. \tag{2.51}$$

As stated in Sect. 2.2.5, the detected EPR signal V consists of a real and an imaginary part, $V = -M_y + iM_x$. Extending the Bloch equations, one can write

$$\frac{dV}{dt} = -\frac{dM_y}{dt} + i\frac{dM_x}{dt} = -V\left[\frac{1}{T_2} + i\Omega_\mathrm{S}\right] + \omega_1 M_0. \tag{2.52}$$

In the present case, each species j gives rise to an EPR transition with a resonance frequency $\omega + \omega_j$. The contributions to the total magnetization from each site can be expressed as

$$\frac{dV_j}{dt} = -V\left[\frac{1}{T_{2,j}} + i(\Omega_\mathrm{S} + \omega_j)\right] + \omega_1 M_{0,j}. \tag{2.53}$$

One can further write $M_{0,j} = \rho_j M_0$ and $W_j = \Omega_\mathrm{S} + \omega_j$. Using the short notation $T_j = T_{2,j}$ and adding the appropriate terms for the transfer of spins between the two sides, Eqs. 2.54, 2.55 are obtained [35].

$$\frac{dV_\mathrm{A}}{dt} = -V\left[\frac{1}{T_\mathrm{A}} + i(\Omega_\mathrm{S} + \omega_\mathrm{A})\right] + \rho_\mathrm{A}\omega_1 M_0 + \frac{V_\mathrm{B}}{\tau_\mathrm{B}} - \frac{V_\mathrm{A}}{\tau_\mathrm{A}} \tag{2.54}$$

$$\frac{dV_B}{dt} = -V\left[\frac{1}{T_B} + i(\Omega_S + \omega_B)\right] + \rho_B\omega_1 M_0 + \frac{V_A}{\tau_A} - \frac{V_B}{\tau_B} \tag{2.55}$$

In equilibrium, the time derivatives can be set to zero. The real part of the magnetization is given by

$$-M_y = -M_{y,A} - M_{y,B}. \tag{2.56}$$

For the two-site problem under consideration, one yields the analytical expression

$$-M_y = \omega_1 M_0 \frac{\begin{array}{c}\left\{\tau_A + \tau_B + \tau_A\tau_B(\rho_A T_B^{-1} + \rho_B T_A^{-1})\right\}\left\{\begin{array}{c}\tau_A\tau_B(T_A^{-1}T_B^{-1} - W_A W_B)\\ +\tau_A T_A^{-1} + \tau_B T_B^{-1}\end{array}\right\}\\ +\tau_A\tau_B(\rho_A W_B + \rho_B W_A)\left\{\tau_A W_A(1 + \tau_B T_B^{-1}) + \tau_B W_B(1 + \tau_A T_A^{-1})\right\}\end{array}}{\begin{array}{c}\left\{\tau_A\tau_B(T_A^{-1}T_B^{-1} - W_A W_B) + \tau_A T_A^{-1} + \tau_B T_B^{-1}\right\}^2\\ +\left\{\tau_A W_A(1 + \tau_B T_B^{-1}) + \tau_B W_B(1 + \tau_A T_A^{-1})\right\}^2\end{array}}. \tag{2.57}$$

A simplification of Eq. 2.57 can be achieved by applying limiting conditions. In the limit of slow exchange, i.e. $\tau_j^{-1} \ll \Delta\omega = |\omega_A - \omega_B|$, the signal consists of two Lorentzian lines, centered at ω_A and ω_B, with line widths T_A^{-1} and T_B^{-1} and relative intensities ρ_A and ρ_B, respectively,

$$-M_y = \omega_1 M_0\left[\rho_A\frac{T_A^{-1}}{T_A^{-2} + W_A^2} + \rho_B\frac{T_B^{-1}}{T_B^{-2} + W_B^2}\right]. \tag{2.58}$$

As the lifetime decreases, but still in the slow exchange regime, each line receives an additional contribution to its line width $T_{exch}^{-1} = \tau_j^{-1}$. Upon further decrease of the lifetime, the two lines merge to a single broadened Lorentzian line at the weighted frequency $\rho_A\omega_A + \rho_B\omega_B$, when $\tau_j^{-1} > \Delta\omega$.

In this condition of fast exchange, the result assumes the simplified form

$$-M_y = \omega_1 M_0 \frac{\rho_A T_A^{-1} + \rho_B T_B^{-1} + \rho_A^2\rho_B^2(\Delta\omega)^2(\tau_A + \tau_B)}{\left[\rho_A T_A^{-1} + \rho_B T_B^{-1} + \rho_A^2\rho_B^2(\Delta\omega)^2(\tau_A + \tau_B)\right]^2 + (\rho_A W_A + \rho_B W_B)^2}. \tag{2.59}$$

Defining a reduced lifetime τ by

$$2\tau^{-1} = \tau_A^{-1} + \tau_B^{-1} \tag{2.60}$$

and using Eq. 2.51, the exchange contribution in the condition of fast exchange amounts to

$$T_{exch}^{-1} = \frac{1}{2}\rho_A\rho_B(\Delta\omega)^2\tau. \tag{2.61}$$

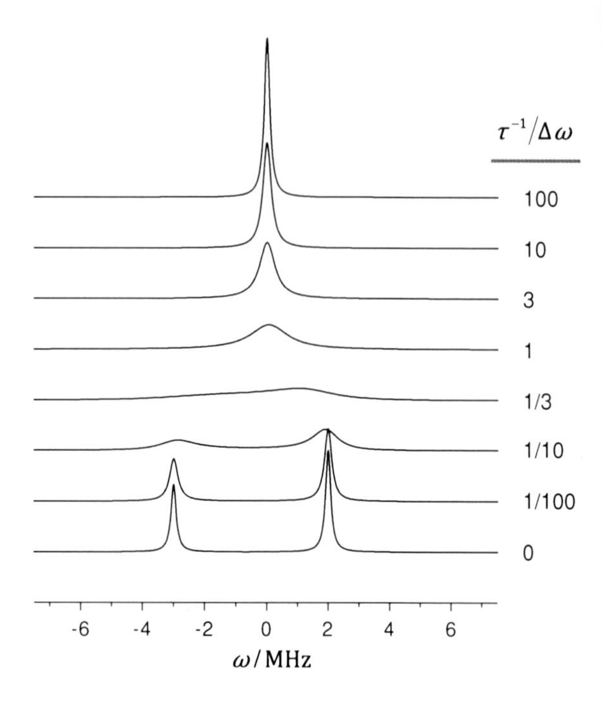

Fig. 2.8 Calculated absorption spectra of a two-spin system with resonance frequencies $\omega_A = 2$ MHz and $\omega_B = -3$ MHz, Lorentzian line widths $T_j^{-1} = 0.1$ MHz, and the relative lifetimes $\tau_A/\tau_B = 3/2$ under evolution of dynamic exchange. As the reduced lifetime τ approaches the frequency difference $\Delta\omega$, the two lines centered at ω_A and ω_B broaden and merge to a single broadened line, which sharpens again at still shorter lifetimes τ

At still shorter lifetimes, the line width becomes independent of the exchange rate and the spectrum consists of a single sharp Lorentzian line with a weighted resonance frequency and line width. The evolution of an EPR spectrum under dynamic exchange is visualized in Fig. 2.8.

2.6 Nitroxides as Spin Probes

As mentioned in Sect. 2.2.1, one major drawback of EPR spectroscopy is related to the small number of paramagnetic systems. The structural determination of these systems by EPR can nonetheless be achieved by introduction of paramagnetic tracer molecules, so-called spin probes or spin labels. Nitroxides, stable free radicals with the structural unit R_2NO, mark the by far most important class of these paramagnetic molecules. The chemical structures of common nitroxide derivatives are shown in Fig. 2.9. In this thesis, nitroxides with structural motives **1** and **4** are used.

The stability of nitroxides can be explained by the delocalized spin density ($\sim$40% on the nitrogen atom, $\sim$60% on the oxygen atom) and full methyl substitution in β-position. The latter provides sterical hindrance as well as removal of the structural motif of a β-proton. Thus, typical radical reactions like dimerizations and disproportionations are inhibited.

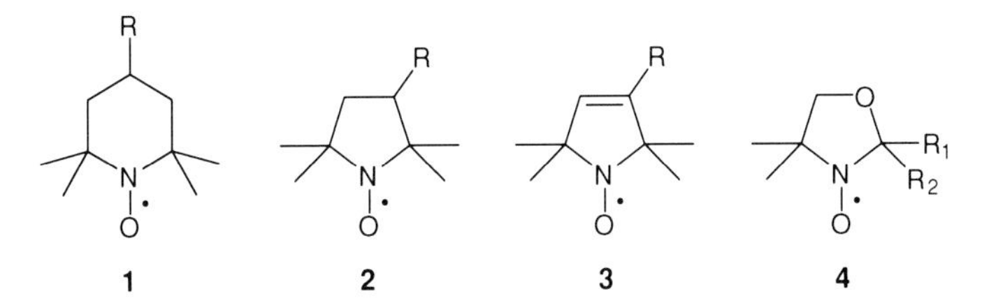

Fig. 2.9 Chemical structures of common nitroxide spin probes. Depicted are derivatives of **a** 2,2,6,6-tetramethylpiperidine-1-oxyl (TEMPO), **b** 2,2,5,5-tetramethylpyrrolidine-1-oxyl (PROXYL), **c** 2,2,5,5-tetramethylpyrroline-1-oxyl (dehydro-PROXYL), and **d** 4,4-dimethyl-oxazolidine-1-oxyl (DOXYL)

2.6.1 Spin Probe versus Spin Label

Spin labels are chemically attached to the material of interest while spin probes interact non-covalently with the system. The method of introduction depends on the system itself and on the type of requested information.

Spin probes are favorable for studies of macromolecules and supramolecular assemblies, as their structure and dynamics are mostly governed by non-covalent interactions. Via self-assembly, the spin probes selectively choose the environment that offers the most attractive interactions. They can also be seen as tracers for small guest molecules like drugs or analytes and deliver information about the host–guest relationship.

The introduction of spin labels is more demanding as the structural unit bearing the unpaired electron needs to be chemically attached to a functional group of the system. This method implies a bigger modification of the system, which could alter its function. It can still be worth the effort and risk, as the electron spin is locally restricted to a specific site of the system and distance measurements can be applied to gain structural information of its three-dimensional structure.

In the last few years, this so-called site directed spin labeling in combination with distance measurements has become a widely used tool for the structural determination of proteins and other (biological) macromolecules [23, 36–38].

2.6.2 Quantum Mechanical Description

For equally distributed nitroxides in dilute solution ($c < 2$ mM), many terms of the static spin Hamiltonian (Eq. 2.20) are irrelevant. In this case, the interactions of the unpaired electron and the magnetic ^{14}N nucleus ($I = 1$) are sufficiently characterized by the electron and nuclear Zeeman interactions and by the hyperfine coupling term, yielding a spin Hamiltonian

$$\mathcal{H}_{\mathrm{NO}} = \beta_{\mathrm{e}} \mathbf{B}_0^{\mathrm{T}} \mathbf{g} \mathbf{S} - \frac{\beta_{\mathrm{N}} g_{\mathrm{N}}}{\hbar} \mathbf{B}_0^{\mathrm{T}} \mathbf{I} + \mathbf{S}^{\mathrm{T}} \mathbf{A} \mathbf{I}. \tag{2.62}$$

The quadrupolar contribution of the ^{14}N nucleus is neglected in this description. The energy eigenvalues are determined through solution of the Schrödinger equation. For a fast, isotropic rotation, six eigenstates with energies

$$E_{\mathrm{NO}} = g_{\mathrm{iso}}\beta_{\mathrm{e}}B_0 m_{\mathrm{S}} - g_{\mathrm{N}}\beta_{\mathrm{N}}B_0 m_{\mathrm{I}} + a_{\mathrm{iso}}m_{\mathrm{S}}m_{\mathrm{I}} \tag{2.63}$$

are obtained. This equation also holds for anisotropic motion, if g_{iso} and a_{iso} are substituted by the effective values in the respective orientation. The corresponding energy level diagram is depicted in Fig. 2.10. Three allowed EPR transitions with frequencies ω_{I} are determined by the selection rules $\Delta m_{\mathrm{S}} = \pm 1$ and $\Delta m_{\mathrm{I}} = 0$,

$$\Delta E_{NO} = \hbar\omega_{\mathrm{I}} = g_{\mathrm{iso}}\beta_{\mathrm{e}}B_0 + a_{\mathrm{iso}}m_{\mathrm{I}}. \tag{2.64}$$

Although the nuclear Zeeman interaction does not affect the EPR transition frequencies, the nuclear frequencies, detectable by advanced EPR methods, depend on the interplay of nuclear Zeeman interaction and hyperfine coupling. A detailed description is given in Sect. 2.10, when the corresponding method is introduced.

The molecular coordinate system of nitroxides is presented in Fig. 2.11. The $2p_z$ orbital of the nitrogen atom defines the z-axis. The x-axis is directed along the N–O bond and the y-axis is directed perpendicular to the xz-plane. If spectra were recorded that only detected orientations along the principal axes, they would assume the form illustrated in Fig. 2.11. The center line position marks the principal **g** tensor element, and the spacing is determined by the corresponding principal element of the hyperfine tensor. The maximum hyperfine coupling value is found along z, since most spin density resides in the π^* orbital along this axis.

2.6.3 Nitroxide Dynamics

The motion of the spin probe is strongly influenced by the dynamics and local structure of its surrounding. Thus, the analysis of nitroxide dynamics delivers indirect information about the system of interest. EPR is sensitive to rotational diffusion rather than translational motion, since only angular motions relative to the external magnetic field affect the magnetic interactions and the spectral line shapes. If the spin probe rotates very fast, the hyperfine couplings in different directions average to an isotropic value a_{iso} and a spectrum is obtained that consists of three equal lines. On the other hand, an anisotropic powder spectrum is obtained when rotational motion is prohibited (cf. Sect. 2.4).

Between these two extremes, the spectral shape strongly depends on the time frame of the rotational diffusion. For isotropic Brownian rotational motion, it is characterized by the rotational correlation time τ_c. Formally, τ_c is calculated by summation of all autocorrelation functions of the Wigner rotation matrices $\mathbf{D}(\Omega(t))$,

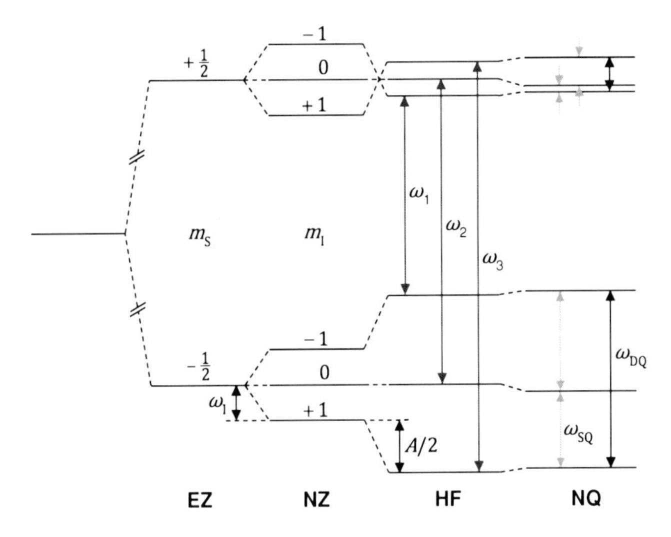

Fig. 2.10 Energy level diagram for an $S = 1/2$, $I = 1/2$ spin system in the strong coupling case $|A/2| > |\Omega_I|$ with allowed EPR transitions (*blue*) and nuclear single quantum (SQ) and double quantum (DQ) transitions (*red*). The nuclear quadrupole contribution is included

$$\tau_c = \int_{t_0}^{\infty} \langle D^l_{m,n}(\Omega(t)) | D^l_{m,n}(\Omega(t_0)) \rangle dt. \tag{2.65}$$

A detailed derivation is given in the literature [40, 41]. The rotational correlation time is related to the rotational diffusion coefficient by

$$\tau_c = \frac{1}{6D_R}. \tag{2.66}$$

The effect of the rotational diffusion on EPR spectra is shown in Fig. 2.12. In the regime of fast rotation (left hand side), the relative widths and heights of the spectral lines change due to increasing anisotropic contributions. The high-field line is most strongly affected since it experiences the cumulative effect of **g** and **A** anisotropy. The strongest spectral changes are observed at $\tau_c \sim 3$ ns, when the inverse rotational correlation time matches the contribution due to anisotropic line broadening. When the rotational motion is further restricted (slow motion regime), the separation of the outer extrema is a meaningful parameter to describe the spin probe dynamics [42].

Though there are established methods to infer the rotational correlation time in the fast motion regime from the relative heights of the central and high-field line [43, 44], these methods break down if the spectra consist of several overlapping species. Due to this fact, all rotational correlation times in this thesis were obtained by fitting of spectral simulations (cf. Sect. 2.7).

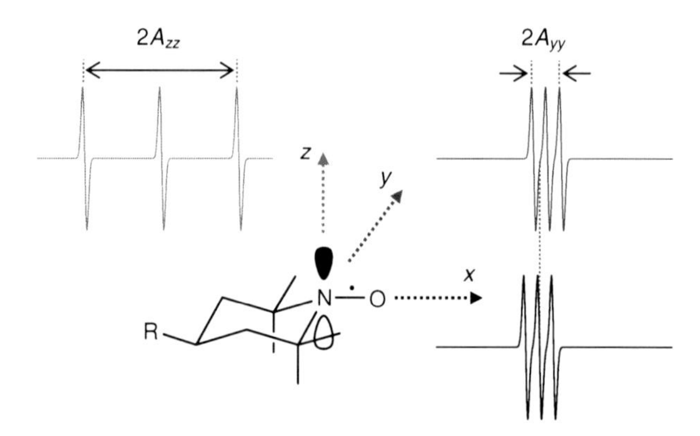

Fig. 2.11 Definition of the molecular coordinate system of nitroxides and hypothetical spectra for orientations along the principal axes. Collinear **g** and **A** tensors with principal values $g_{xx} = 2.0087$, $g_{yy} = 2.0065$, $g_{zz} = 2.0023$ and $A_{xx} = A_{yy} = 18.2$ MHz, $A_{zz} = 100.9$ MHz were assumed. Adapted from [39] with permission from the author

If the rotation of the nitroxide is hindered in several directions (e.g. due to chemical attachment), an anisotropy has to be introduced to the Brownian motion. In case of a fast rotation about one axis and a slower rotation about directions perpendicular to that axis, the rotational diffusion coefficient can be expressed in its principal axes frame by [40, 45]

$$\mathbf{D}_{\mathrm{R}} = \begin{pmatrix} D_{\perp} & & \\ & D_{\perp} & \\ & & D_{\parallel} \end{pmatrix}. \tag{2.67}$$

Examples for more sophisticated approaches to account for spin probe dynamics are the macroscopic order microscopic disorder (MOMD) model and the slowly relaxing local structure (SRLS) model [46, 47]. They are not introduced in detail, since in this work it was sufficient to analyze the spin probe rotation in terms of isotropic and anisotropic rotational motion.

2.6.4 Environmental Influences

In addition to dynamics, spin probes also provide information about their local environment. The electronic structure of a nitroxide is slightly altered depending on the interactions with molecules in its surrounding. In solvents of different polarity and proticity (pH), the same spin probe thus shows slightly different hyperfine coupling constants and g-values. For X-band spectra, this effect is illustrated in Fig. 2.13. The alterations of the spectra are small but add up at the high-field line. Thus, spin probes in different nanoscopic environments in inhomogeneous samples can be distinguished and analyzed separately. This constitutes

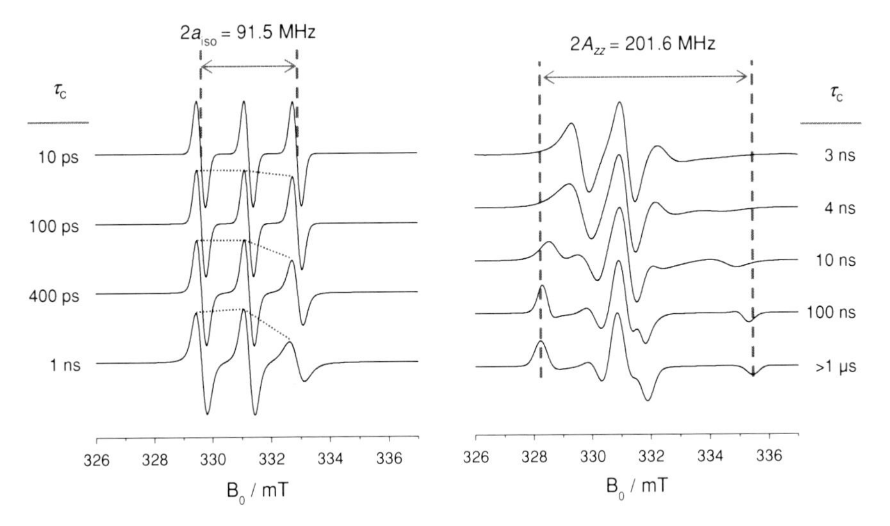

Fig. 2.12 EPR spectral dependence on rotational dynamics for isotropic rotational diffusion in the regimes of fast motion (*left*) and slow tumbling (*right*). The spectra were obtained by simulations with routines accounting for fast motion $\tau_c = (10–400\ \text{ps})$, intermediate/slow motion $\tau_c = (1–100\text{ns})$, and powder type spectra in the rigid limit ($\tau_c > 1\,\mu\text{s}$) (cf. Sect. 2.7). The principal values of **g** and **A** are listed in Fig. 2.11. Adapted from [79] with permission from the author

the main source of information for the characterization of thermoresponsive systems in Chaps. 5 and 7.

In polar solvents, the zwitterionic resonance structure in Fig. 2.14b is stabilized and spin density is transferred to the nitrogen nucleus. This leads to an increase of A_{zz} and to a decrease of g_{xx} since the deviation of g_{xx} from g_e mainly depends on the spin–orbit coupling in the oxygen orbitals.

At a given polarity, the SOMO–LUMO energy difference increases, if the oxygen atom acts as hydrogen acceptor (Fig. 2.14a). Hence, the spin–orbit contribution decreases, and the deviation of g_{xx} from g_e becomes less significant (cf. Sect. 2.3.1). In high-field spectra, the correlation of g_{xx} and A_{zz} can be utilized to distinguish between polar and protic solvents [48, 49]. At X-band and for fast rotating spin probes, the above-mentioned effects manifest themselves in changes of a_{iso} and g_{iso}.

2.7 CW Spectral Analysis via Simulations

Only for simple CW EPR spectra, all structural and dynamic parameters can be inferred by a straightforward analysis. This method breaks down in case of overlapping spectral components, complex rotational dynamics, and unresolved couplings. Here, much information is buried in the line shape and cannot be retrieved in a non-trivial manner. In this case, the reproduction of the spectrum by

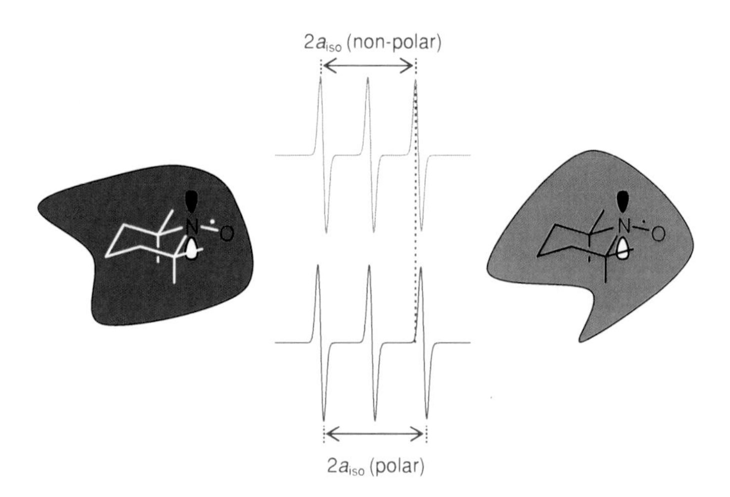

Fig. 2.13 Schematic depiction of spin probes in hydrophilic (*blue*) and hydrophobic (*orange*) environments and corresponding fast-motion spectra with g_{iso} (polar) = 2.0060, a_{iso} (polar) = 48.2 MHz and g_{iso} (non-polar) = 2.0063, a_{iso} (non-polar) = 44.3 MHz

a quantum-mechanical simulation is the method of choice. The spectral parameters of interest can then be read out from the simulation.

Most CW EPR spectra in this thesis were analyzed by spectral simulations, performed with home-written Matlab programs based on various EasySpin routines. EasySpin is a simulation software developed by Stoll and Schweiger that provides routines for spectral simulations in a wide range of dynamic conditions [50, 51]. It subdivides four regimes of rotational motion: the isotropic limit, fast motion, slow motion, and the rigid limit.

The fast tumbling of radicals in the isotropic limit leads to a complete averaging of rotational motion and the spectrum consists of symmetric lines with equal widths. Corresponding spectra are computed by a diagonalization of the isotropic spin Hamiltonian. The resulting energy levels are obtained by the Breit-Rabi formulae [52].

In the fast motion regime (e.g. Fig. 2.12 left), small anisotropy effects change the heights and widths of the EPR transitions depending on their nuclear quantum number m_I. Applying the Redfield theory, these effects are treated as small perturbation [53, 54].

In slow motion (e.g. Fig. 2.12 right), the perturbation approach does not sufficiently account for the large spectral influence of rotational motion. Spectra are calculated by solving the stochastic Liouville equation in an approach developed by Schneider and Freed [45]. The rotational motion is accounted for by the diffusion superoperator, which depends on the model and the orientational distribution of rotation.

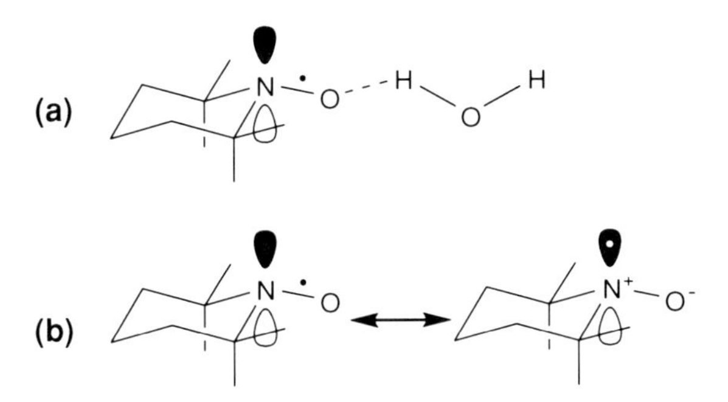

Fig. 2.14 Origin of the polarity and proticity dependence of nitroxide spectra. **a** Hydrogen bonding leads to a decrease of the spin–orbit coupling. **b** A stabilization of the zwitterionic resonance structure causes an increase of the spin density at the nitrogen atom. The combined effects lead to a decrease of g_{iso} and to an increase of a_{iso}

In the rigid limit, the orientations of the paramagnetic molecules are static. In this case, the positions, intensities and widths of all resonance lines are computed for each orientation by modeling the energy level diagram [55].

2.8 Time Evolution of Spin Ensembles

Although a CW EPR spectrum is influenced by all magnetic interactions of the spin Hamiltonian, only few dominating interactions are sufficiently resolved for an accurate analysis. The different contributions to the spin Hamiltonian can be separated by pulse EPR spectroscopy. Thus, small interactions of the electron spin with remote electron and nuclear spins can be resolved and precisely measured. Since in pulse EPR spectroscopy, microwave irradiation is not applied continuously but in defined time intervals, the time evolution of the spins has to be considered explicitly.

In this section, a formalism is introduced that describes the quantum-mechanical state of spin ensembles. Further, it is shown how this formalism can be applied to EPR. Based on this foundation, those pulse methods applied in this thesis are introduced in the following sections.

2.8.1 *The Density Matrix*

The quantum-mechanical expectation value of an observable O in a (single) spin system described by the wave function ψ is given by

$$\langle O \rangle = \langle \psi | O | \psi \rangle. \tag{2.68}$$

In this representation, the explicit form of the wave function has to be known. The wave function can be developed in a set of orthogonal eigenfunctions k

$$|\psi(t)\rangle = \sum_k c_k(t)|k\rangle, \tag{2.69}$$

yielding a new expression for the expectation value

$$\langle O\rangle = \sum_{k,l} \overline{c_k c_l^*} \langle l|O|k\rangle. \tag{2.70}$$

In reality, a spin ensemble is detected rather than a single spin. Their properties are best described by the Hermitian density operator

$$\boldsymbol{\rho} = |\psi\rangle\langle\psi| = \sum_{k,l} \overline{c_k c_l^*} |k\rangle\langle l| \tag{2.71}$$

with the matrix elements

$$\rho_{kl} = \overline{c_k c_l^*} = \langle k|\boldsymbol{\rho}|l\rangle. \tag{2.72}$$

The off-diagonal matrix element ρ_{kl} quantifies coherences between states $|k\rangle$ and $|l\rangle$, while the diagonal elements ρ_{kk} represent the population of the state $|k\rangle$. Ensemble averaging is done by an averaging of the coefficients c_k, c_l. The expectation value of an observable in an ensemble is then given by

$$\langle\overline{O}\rangle = \mathrm{Tr}(\boldsymbol{\rho}\mathbf{O}), \tag{2.73}$$

which is conveniently independent of the wave function. In analogy to the time dependent Schrödinger equation

$$\frac{d|\psi\rangle}{dt} = -i\mathcal{H}|\psi(t)\rangle, \tag{2.74}$$

the evolution of the density matrix under the spin Hamiltonian is given by the Liouville–von Neumann equation

$$\frac{d|\boldsymbol{\rho}\rangle}{dt} = -i[\mathcal{H}|\boldsymbol{\rho}]. \tag{2.75}$$

The Liouville–von Neumann equation neglects relaxation which can be introduced by the relaxation super-operator Ξ to yield

$$\frac{d|\boldsymbol{\rho}\rangle}{dt} = -i[\mathcal{H}|\boldsymbol{\rho}] - \Xi\left[\boldsymbol{\rho}(t) - \boldsymbol{\rho}_{\mathrm{eq}}\right]. \tag{2.76}$$

where $\boldsymbol{\rho}_{\mathrm{eq}}$ is the density operator of a spin system in thermal equilibrium. Since Ξ is time-independent, relaxation and time evolution of the spin ensemble can be treated separately.

The Liouville–von Neumann equation is readily solved when the Hamiltonian is constant for some time. The solution is most easily written in terms of operator exponentials

$$\boldsymbol{\rho}(t) = \exp(-i\mathcal{H}t)\boldsymbol{\rho}(0)\exp(-i\mathcal{H}t) = \mathbf{U}\boldsymbol{\rho}(0)\mathbf{U}^{\dagger}. \tag{2.77}$$

The propagator operator $\mathbf{U}$ and its Hermitian adjunct $\mathbf{U}^{\dagger}$ propagate the density operator in time via unitary transformations.

Solving the Liouville–von Neumann equation is more difficult when the Hamiltonian varies with time, e.g. in an experiment with pulsed microwave irradiation. In this case, it is convenient to divide the experiment in time intervals, during which the Hamiltonian is constant. In the framework of this so-called density operator formalism, propagators are consecutively applied for each time interval

$$\boldsymbol{\rho}(t_n) = \mathbf{U}_n \cdots \mathbf{U}_2\mathbf{U}_1\boldsymbol{\rho}(0)\mathbf{U}_1^{\dagger}\mathbf{U}_2^{\dagger} \cdots \mathbf{U}_n^{\dagger}. \tag{2.78}$$

2.8.2 Product Operator Formalism

For large spin systems, it is more convenient to decompose ρ into a linear combination of orthogonal basis operators B_j

$$\boldsymbol{\rho}(t) = \sum_j b_j(t)\mathbf{B}_j. \tag{2.79}$$

This product operator formalism is advantageous since each of the basis operators has a certain physical meaning [56]. Each set of basis operators consists of as many operators as there are elements in the density matrix. They span a basis with dimension $n_{\mathrm{H}}^2 = n_{\mathrm{L}}$, called Liouville space. For a single electron spin with $S = 1/2$, a Cartesian basis of $\mathbf{S}_x$, $\mathbf{S}_y$, $\mathbf{S}_z$ and half the identity operator $\frac{1}{2}\mathbf{1}$ can be used.

Not only density operators, but also Hamiltonians can be expressed in terms of Cartesian product operators. Thus, the evolution of the spin system can be described in the framework of product operators. In analogy to Eq. 2.77, the evolution of any operator A under another operator B can be expressed by

$$\begin{aligned} \exp(-i\phi B)A\exp(i\phi B) &= C \\ &\equiv A \xrightarrow{\phi B} C \end{aligned} \tag{2.80}$$

It is convenient to use the shorthand notation introduced in the lower part of eq. 2.80 to describe real EPR experiments. A generalized solution for $J = 1/2$ spin systems is given by [57]

$$A \xrightarrow{\phi B} A\cos\phi - i[B,A]\sin\phi \quad \forall\,[B,A] \neq 0,$$
$$A \xrightarrow{\phi B} A \quad \forall\,[B,A] = 0. \tag{2.81}$$

If the Hamiltonian consists of a sum of product operator terms, they can be applied consecutively as long as they commute with each other.

2.8.3 Application to EPR

Before manipulating a spin system by microwave irradiation, its initial state must be known. The equilibrium state is characterized by a Boltzmann distribution (see also Eq. 2.4), i.e. the equilibrium density operator $\boldsymbol{\rho}_{\text{eq}}$ is given by

$$\boldsymbol{\rho}_{\text{eq}} = \frac{\exp(-\mathcal{H}_0\hbar/kT)}{\text{Tr}[\exp(-\mathcal{H}_0\hbar/kT)]}. \tag{2.82}$$

In the high-field approximation, the dominating interaction is given by the electron Zeeman term and $\mathcal{H}_0 \cong \omega_S \mathbf{S}_z$. Using the high-temperature approximation $\omega_S \mathbf{S}_z \gg kT$ and performing a series expansion of the exponential, $\boldsymbol{\rho}_{\text{eq}}$ can be approximated by

$$\boldsymbol{\rho}_{\text{eq}} \cong \mathbf{1} - \frac{\hbar\omega_S}{kT}\mathbf{S}_z. \tag{2.83}$$

The identity operator $\mathbf{1}$ is usually neglected as it is invariant throughout the experiment. Further, the constant factors are dropped and the equilibrium density operator is conveniently denoted as

$$\boldsymbol{\rho}_{\text{eq}} = -\mathbf{S}_z. \tag{2.84}$$

The spin system leaves the state of thermal equilibrium by application of microwave pulses, i.e. (resonant) microwave irradiation during a time interval t_P in which the magnetization of the spin system is rotated by a flip angle

$$\beta = \omega_1 t_P. \tag{2.85}$$

A pulse excites one or several transitions of a spin system depending on its excitation bandwidth, which is inversely proportional to t_P. A pulse is called non-selective, if it excites all allowed and partially allowed transitions on the spin system, or selective, if only one (or few) transitions are excited. Quantum-mechanically, a non-selective pulse applied from direction $i \in \{x, y, -x, -y\}$ is described by the product operator $\beta\mathbf{S}_i$. In this case, the static spin Hamiltonian during the pulse can be neglected and the pulse can be treated as an ideal, infinitely short pulse.

The spin Hamiltonian does, however, affect the further evolution of the spin system, as described in the Liouville–von Neumann equation (Eq. 2.77). The time of free evolution is thus characterized by the operator $\mathcal{H}_0 t$, or its linear combination of product operators.

In conclusion, starting from the point of thermal equilibrium, the fate of the spin system at the time of detection can be determined by consecutive application of product operators describing microwave pulses and free evolution.

2.8.4 The Vector Model

The vector model offers a somewhat more intuitive classical description for pulse EPR experiments, since the fate of the magnetization is followed in a pictorial view. The magnetization vector **M** was already introduced in Sect. 2.2.3 in the context of the Bloch equations (cf. Fig. 2.2). In analogy to the quantum-mechanical treatment, a pulse rotates **M** about the direction of the microwave field i by the flip angle β. In a shorthand notation, the pulse is referred to as β_i-pulse. Note that due to the sign convention of Eq. 2.1, the magnetization vector is aligned antiparallel to **S**.

The vector model is severely limited by the fact that it only applies to uncoupled spins, but it offers a conceptional insight in some key experiments. In the following chapters, it is used to illustrate the formation of spin echoes in addition to the quantum-mechanical treatment.

2.8.5 The ($S = 1/2, I = 1/2$) Model System

The principles of many pulse EPR experiments can be explained by considering a spin system consisting of one electron spin with $S = 1/2$ which is coupled to one nuclear spin with $I = 1/2$. The four-level scheme of such a system is depicted in Fig. 2.15a.

Since there is a certain probability for the nuclear spin to be flipped by a microwave pulse, also forbidden double and zero-quantum transitions are excited to some extent. Thus, both allowed and forbidden transitions with frequencies ω_{kl} give rise to an EPR spectrum illustrated in Fig. 2.15b.

The allowedness of a transition is characterized by $\eta = (\eta_\alpha - \eta_\beta)/2$. η_α and η_β define the directions of the nuclear quantization axes in the two m_S manifolds with respect to $\mathbf{B}_0$. Allowed transitions are weighted with $\cos\eta$, forbidden transitions with $\sin\eta$. The redistribution of coherences due to the excitation of forbidden transitions is called branching.

The two-spin system is fully described by the density matrix given in Fig. 2.15c. The diagonal elements ρ_{kk} represent populations of the corresponding

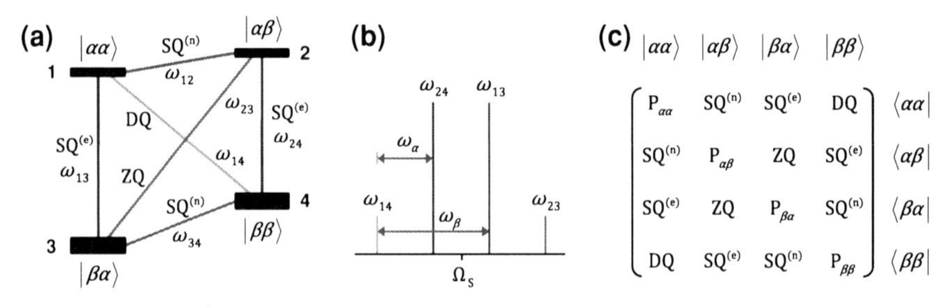

Fig. 2.15 Description of a weakly coupled $S = 1/2$, $I = 1/2$ spin system ($|\omega_I| > |A/2|$). **a** Energy level diagram with allowed single quantum (SQ) transitions of the electron (*blue*) and nucleus (*red*) and forbidden EPR double quantum (DQ) and zero quantum (ZQ) transitions (*orange/green*). **b** Corresponding stick spectrum. **c** Density matrix. The diagonal elements ρ_{kk} represent populations of the corresponding spin states, the off-diagonal elements ρ_{kl} represent coherences. Reproduced from [76] by permission of Oxford University Press (www.uop.com)

spin states, the off-diagonal elements ρ_{kl} represent coherences. The Cartesian basis of the two-spin system is obtained by multiplication of the Cartesian basis sets of the single spins.

$$\begin{aligned}\{\mathbf{A}_1, \mathbf{A}_2, \cdots, \mathbf{A}_{16}\} &= 2 \times \left\{\frac{1}{2}\mathbf{1}, \mathbf{S}_x, \mathbf{S}_y, \mathbf{S}_z\right\} \otimes \left\{\frac{1}{2}\mathbf{1}, \mathbf{I}_x, \mathbf{I}_y, \mathbf{I}_z\right\} \\ &= \left\{\begin{matrix} \frac{1}{2}\mathbf{1}, \mathbf{S}_x, \mathbf{S}_y, \mathbf{S}_z, \mathbf{I}_x, \mathbf{I}_y, \mathbf{I}_z, 2\mathbf{S}_x\mathbf{I}_x, \\ 2\mathbf{S}_x\mathbf{I}_y, 2\mathbf{S}_x\mathbf{I}_z, 2\mathbf{S}_y\mathbf{I}_x, 2\mathbf{S}_y\mathbf{I}_y, 2\mathbf{S}_y\mathbf{I}_z, 2\mathbf{S}_z\mathbf{I}_x, 2\mathbf{S}_z\mathbf{I}_y, 2\mathbf{S}_z\mathbf{I}_z \end{matrix}\right\}\end{aligned} \tag{2.86}$$

The Hamiltonian of this spin system in its eigenbasis is given by

$$\mathcal{H}_0 = \Omega_S\mathbf{S}_z + \frac{\omega_+}{2}\mathbf{I}_z + \omega_-\mathbf{S}_z\mathbf{I}_z \tag{2.87}$$

with the nuclear combination frequencies

$$\begin{aligned}\omega_+ &= \omega_{12} + \omega_{34}, \\ \omega_- &= \omega_{12} - \omega_{34},\end{aligned} \tag{2.88}$$

as defined in Fig. 2.15b. Depending on the signs and magnitudes of the hyperfine coupling and nuclear Zeeman interaction, the nuclear frequencies ω_{12} and ω_{34} can assume positive or negative values. In some cases, it is convenient to use absolute values, defined by $\omega_\alpha = |\omega_{12}|$ and $\omega_\beta = |\omega_{34}|$.

2.9 Pulse EPR Methods Based on the Primary Echo

The standard experiment in NMR spectroscopy consists of a single $\pi/2$ pulse, which excites the whole spin system. Then, the free induction decay of the generated nuclear coherences is detected and the time-domain signal is Fourier transformed. The corresponding FT EPR technique is rarely used as it suffers from major problems. First, the excitation bandwidth of the shortest available microwave pulses ($t_{\pi/2} \approx 4$ ns) does not allow for a complete excitation of spin systems with even moderately broad spectral widths. More severely, the strong excitation pulse causes a ringing of the resonator, which prevents the recording of a signal for a certain time after the pulse. This so-called deadtime is $\sim$100 ns at X-band.

To circumvent the deadtime problem, most EPR experiments are based on the detection of electron spin echoes (ESE). The spin echo, invented by Hahn in 1950 [58], describes the reappearance of magnetization as an echo of the initial magnetization. The simplest representative of a spin echo, the primary echo (or Hahn echo), is observed after the pulse sequence $\pi/2 - \tau - \pi - \tau$, as illustrated in Fig. 2.16.

A vectorial explanation for the formation of the echo is given in Fig. 2.16 (bottom). The first $(\pi/2)_x$-pulse rotates the magnetization vector in the $-y$-direction. During the time of free evolution τ, spin packets with different Larmor frequencies gain a phase shift due to their different angular precession. In the rotating frame, the spin packets thus fan out according to their resonance offset Ω_S and the magnitude of the magnetization vector is decreased. The second $(\pi)_x$-pulse inverts the sign of the y-component. After the same evolution time τ, the phases of the different spin packets are refocused and the magnetization vector is maximized.

The same result is obtained in the framework of the product operator formalism (Eq. 2.81). Using the commutation rules for angular momentum operators

$$\begin{aligned} [\mathbf{S}_x, \mathbf{S}_y] &= i\mathbf{S}_z \\ [\mathbf{S}_y, \mathbf{S}_z] &= i\mathbf{S}_x \\ [\mathbf{S}_z, \mathbf{S}_x] &= i\mathbf{S}_y, \end{aligned} \tag{2.89}$$

the density operators for a system of isolated electron spins with $\mathcal{H}_0 = \Omega_s \mathbf{S}_z$ during a pulse sequence $(\pi/2)_x - \tau - \pi_x - t$ are given by

$$\begin{aligned} \boldsymbol{\rho}_{\mathrm{eq}} = -\mathbf{S}_z \xrightarrow{\pi/2\mathbf{S}_x} \boldsymbol{\rho}_1 = \mathbf{S}_y \xrightarrow{\tau\Omega_s\mathbf{S}_z} \boldsymbol{\rho}_2(\tau) &= \cos(\Omega_s\tau)\mathbf{S}_y - \sin(\Omega_s\tau)\mathbf{S}_x \\ \boldsymbol{\rho}_2(\tau) \xrightarrow{\pi\mathbf{S}_x} \boldsymbol{\rho}_3(\tau) &= -\cos(\Omega_s\tau)\mathbf{S}_y - \sin(\Omega_s\tau)\mathbf{S}_x \\ \boldsymbol{\rho}_3(\tau) \xrightarrow{t\Omega_s\mathbf{S}_z} \boldsymbol{\rho}_4(\tau + t) &= -\cos(\Omega_s(t-\tau))\mathbf{S}_y - \sin(\Omega_s(t-\tau))\mathbf{S}_x. \end{aligned} \tag{2.90}$$

For $\tau = t$, the arguments of the sin and cos functions vanish and the density operator $\boldsymbol{\rho}_4$ becomes

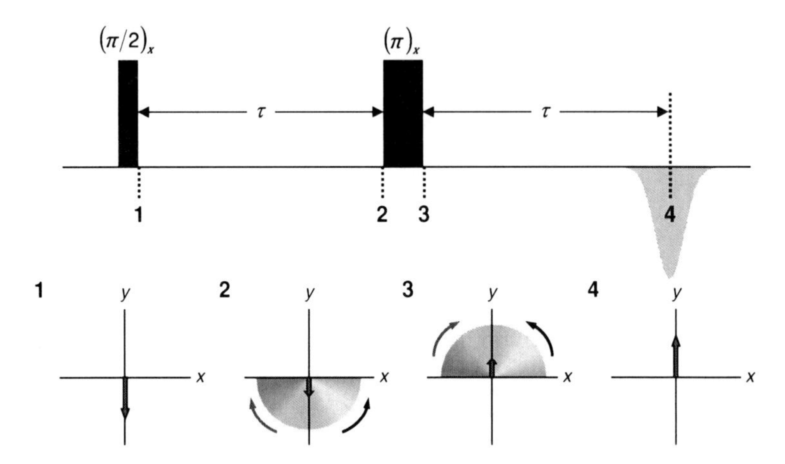

Fig. 2.16 *Top*: Formation of the primary echo after the pulse sequence $(\pi/2)_x - \tau - (\pi)_x - \tau$. *Bottom*: Graphical illustration of the magnetization vector at characteristic positions in the pulse sequence. Adapted from [79] with permission from the author

$$\boldsymbol{\rho}_4(2\tau) = -\mathbf{S}_y = -\boldsymbol{\rho}_1, \tag{2.91}$$

i.e. the (negative) spin state after the first pulse is retained.

The pulse sequence only refocuses inhomogeneous spin packets. Magnetization that is lost due to relaxation cannot be regained. This gives rise to an exponential decay of the echo as a function of τ. The characteristic time of the exponential decay, the phase memory time T_m, is closely related to the transverse relaxation time T_2.

2.9.1 ESE Detected Spectra

Though the pulse detection circumvents the deadtime problem, it does not overcome the insufficient excitation bandwidth. This problem can be solved by a sweep of the magnetic field analogous to the CW EPR technique. The microwave frequency and the pulse sequence are kept constant and the echo intensity is recorded as a function of the magnetic field. With this method, EPR absorption spectra are obtained.

At ambient temperature, echoes from paramagnetic species are usually not detectable due to short phase memory times which cause a complete relaxation of the signal by the time of detection. In this thesis, ESE-detected spectra were recorded at temperatures that provide for sufficiently long T_m times. Nitroxide spectra were usually recorded at 50 K, samples containing Cu^{2+} were measured at 10–20 K.

2.9.2 2-Pulse Electron Spin Echo Envelope Modulation (ESEEM)

In 1965, Rowan et al. [59] found that the spin echo decay of certain samples is modulated by nuclear frequencies and their combinations. This so-called electron spin envelope modulation (ESEEM) originates from the fact that the second $(\pi)_x$-pulse does not only invert the phase of the electron coherence but also redistributes this coherence among all allowed and forbidden electron spin transitions (cf. Sect. 2.8.5), which precess at different frequencies ω_{kl} [60].

After the time 2τ, only magnetization from the initially inverted electron coherence gives rise to an echo. The redistributed magnetization has acquired a phase ϕ and forms coherence transfer echoes that oscillate with $\cos(\phi)$. For the two-spin system, the phase is given by $\phi = (\omega_{kl} - \omega_{13})\tau$. The oscillations thus depend on the frequencies of the nuclear transitions ω_{12} and ω_{34} and the difference frequency ω_-.

The spin density operator at the time of echo formation can be calculated by substituting the spin Hamiltonian in Eq. 2.90 by that of Eq. 2.87. Thus, the modulation of the electron spin echo is described by

$$V_{2\mathrm{p}}(\tau) = -S_y = 1 - \frac{k}{4}\left[2 - 2\cos(\omega_\alpha \tau) - 2\cos(\omega_\beta \tau) + \cos(\omega_+ \tau) + \cos(\omega_- \tau)\right] \tag{2.92}$$

with a modulation depth parameter

$$k = \sin^2(2\eta) = \frac{9}{4}\left(\frac{\mu_0 g \beta_\mathrm{e}}{4\pi B_0}\right)^2 \frac{\sin^2(2\theta)}{R^6}. \tag{2.93}$$

The last term of Eq. 2.93 is valid for weak hyperfine couplings and an isotropic g-value. In this case, the modulation depth is inversely proportional to the sixth power of the distance between the electron and the nuclear spin R [61].

Modulations of the electron spin echo due to couplings with j nuclei are multiplicative. Further, the modulations are superimposed by the exponential decay of the spin echo. Thus, the experimental time trace amounts to

$$V'_{2\mathrm{p}}(\tau) = \exp\left(\frac{-2\tau}{T_\mathrm{m}}\right)\prod_j V_{2\mathrm{p},j}(\tau). \tag{2.94}$$

To analyze the modulation frequencies, the exponential background is divided out and the time domain data is Fourier transformed into the frequency domain. The apparent resolution of the frequency domain spectrum is improved by zero filling, truncation artefacts are avoided by apodization. These methods apply to the frequency analysis of all ESEEM techniques.

2.10 Pulse EPR Methods Based on the Stimulated Echo

Schemes based on the stimulated echo are among the most commonly applied techniques in pulse EPR spectroscopy. The pulse sequence of a stimulated echo can be viewed as a modified primary echo sequence, in which the second π-pulse is divided in two $\pi/2$-pulses separated by a time T (cf. Fig. 2.17).

Thus, the magnetization vector evolves analogously to the primary echo up to the second pulse, which rotates the magnetization fan about the x-axis. For $T \gg T_{\mathrm{m}}$, the x-components of the rotated fan have vanished due to transverse relaxation. The third $\pi/2$-pulse rotates the remaining z-components of the magnetization vectors in the y-direction. The precession direction and frequency is the same as in the first evolution time τ. Hence, after a second time τ, an echo is formed with the magnetization vectors aligned on a circle centered along the y-axis.

The corresponding quantum-mechanical product operator calculation for a system of isolated spins is described by

$$\boldsymbol{\rho}_{\mathrm{eq}} \xrightarrow{\pi/2\,\mathbf{S}_x} \xrightarrow{\tau\Omega_{\mathrm{s}}\mathbf{S}_z} \xrightarrow{\pi/2\,\mathbf{S}_x} \xrightarrow{T\Omega_{\mathrm{s}}\mathbf{S}_z} \xrightarrow{\pi/2\,\mathbf{S}_x} \xrightarrow{t\Omega_{\mathrm{s}}\mathbf{S}_z} \boldsymbol{\rho}_6(\tau + T + t). \tag{2.95}$$

For $T \gg T_{\mathrm{m}}$, $\boldsymbol{\rho}_6$ amounts to

$$\begin{aligned} \boldsymbol{\rho}_6(\tau + T + t) = &-\frac{1}{2}[\cos(\Omega_{\mathrm{s}}(t-\tau)) + \cos(\Omega_{\mathrm{s}}(t+\tau))]\mathbf{S}_y \\ &+\frac{1}{2}[\sin(\Omega_{\mathrm{s}}(t-\tau)) + \sin(\Omega_{\mathrm{s}}(t+\tau))]\mathbf{S}_x. \end{aligned} \tag{2.96}$$

The terms with arguments $\Omega_{\mathrm{s}}(t-\tau)$ lead to an echo formation at $t = \tau$, the terms with $\Omega_{\mathrm{s}}(t+\tau)$ form a so-called virtual echo at $t = -\tau$, but do not contribute to the stimulated echo. Hence, the stimulated echo is only half as intense as the primary Hahn echo.

In a different description, the sequence $\pi/2 - \tau - \pi/2$ creates a polarization grating (Fig. 2.20), which is stored for the time T. The last $\pi/2$-pulse induces a free induction decay of the grating, which shows the incidental shape of an echo.

2.10.1 3-Pulse ESEEM

In analogy to 2-pulse ESEEM, an envelope modulation of the stimulated echo is observed when the time delay T is incremented. For a two-spin system, the modulation part of the echo signal is given by

$$\begin{aligned} V_{3\mathrm{p}}(\tau, T) = 1 - \frac{k}{4}\{&[1-\cos(\omega_\beta\tau)][1-\cos(\omega_\alpha(\tau+T))] \\ &+[1-\cos(\omega_\alpha\tau)]\left[1-\cos(\omega_\beta(\tau+T))\right]\}. \end{aligned} \tag{2.97}$$

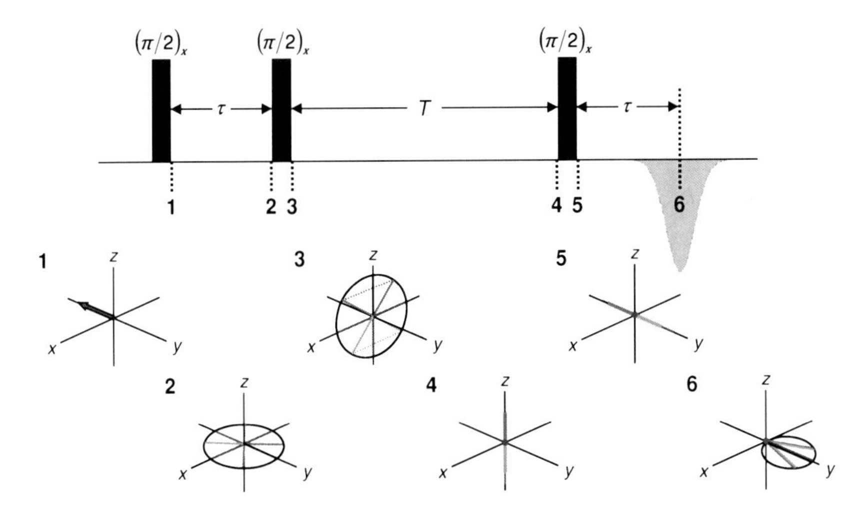

Fig. 2.17 ***Top***: Formation of the stimulated echo after the pulse sequence $(\pi/2)_x - \tau - (\pi/2)_x - T - (\pi/2)_x - \tau$. *Bottom*: Graphical illustration of the magnetization vector (and selected vectors of the magnetization fan, respectively) at characteristic positions in the pulse sequence. Adapted from [79] with permission from the author

Unlike in 2-pulse ESEEM, no nuclear combination frequencies are observed. This significantly simplifies spectral interpretation. Further, an increased spectral resolution is achieved, since the echo decay is related to the phase memory time of the nuclear spins $T_\mathrm{m}^{(\mathrm{n})}$, which is much longer than T_m. The experimental time trace of an electron coupled to a single nuclear spin is given by

$$V'_{3\mathrm{p}}(\tau, T) = \exp\left(\frac{-\tau}{T_\mathrm{m}^{(\mathrm{n})}}\right) V_{3\mathrm{p}}(\tau, T). \tag{2.98}$$

Note that in contrast to 2-pulse ESEEM, modulations due to couplings with several nuclei are only multiplicative with respect to the electron spin manifold.

In Fig. 2.18, schematic nuclear frequency spectra are depicted in the framework of the two-spin system. In case of a negligible anisotropic hyperfine contribution, the modulation frequencies can be calculated by

$$\omega_\alpha = \left|\omega_\mathrm{I} + \frac{A}{2}\right|, \omega_\beta = \left|\omega_\mathrm{I} - \frac{A}{2}\right|. \tag{2.99}$$

Thus, a weak coupling case $|\omega_\mathrm{I}| > |A/2|$ and a strong coupling case $|\omega_\mathrm{I}| < |A/2|$ can be distinguished. In the weak coupling case, the peaks are symmetrically centered around ω_I and separated by A. In the strong coupling case, they are centered around $A/2$ and separated by $2\omega_\mathrm{I}$.

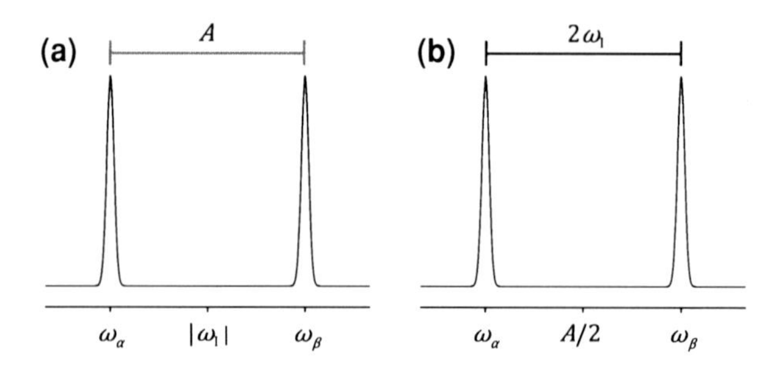

Fig. 2.18 Schematic nuclear frequency spectra for **a** weak coupling ($|\omega_I| > |A/2|$) and **b** strong coupling ($|\omega_I| < |A/2|$) in case of a negligible anisotropic hyperfine contribution. $\omega_I < 0$ and $A > 0$ are assumed, as observed for a proton with $g_n > 0$

2.10.2 Hyperfine Sublevel Correlation (HYSCORE) Spectroscopy

A modification of the 3-pulse ESEEM sequence is achieved by introduction of a non-selective π-pulse, which interchanges the nuclear coherences between the $|\alpha\rangle$ and $|\beta\rangle$ manifolds of the electron spin (Fig. 2.19a) [62]. The incrementation of both times prior to and after this pulse, t_1 and t_2, yields time-domain data in two dimensions which can be converted to 2D nuclear frequency spectra by Fourier transformation. This 2D EPR technique was invented in 1986 by Höfer et al. and named hyperfine sublevel correlation spectroscopy (HYSCORE).

Since spectral density is distributed over two dimensions, HYSCORE spectra provide for an improved resolution and contain information that cannot easily be retrieved by 1D ESEEM methods. The gain in spectral resolution is demonstrated in Fig. 2.19b, where HYSCORE spectra are schematically illustrated for the two-spin system in analogy to the 1D spectra in Fig. 2.18. Peaks of weakly coupled nuclei appear in the first quadrant, the second quadrant contains spectral information about strongly coupled nuclei.

2.10.3 Blind Spots

A feature of the 3-pulse ESEEM and the HYSCORE technique is a consequence of the factors $1 - \cos(\omega_{\alpha,\beta}\tau)$ in Eq. 2.97. For $\omega_{\beta,\alpha} = 2\pi n/\tau$ ($n \in \mathbb{N}_0$) the modulation at frequency $\omega_{\alpha,\beta}$ vanishes. This leads to so-called blind spots in the spectrum where the nuclear frequencies are efficiently suppressed.

The origin of the blindspots can also be described pictorially. The polarization grating created by a $\pi/2 - \tau - \pi/2$ pulse sequence is spaced by $1/\tau$, as illustrated in Fig. 2.20a. Nuclear modulations lead to a polarization transfer that changes the spectral shape and can be read out by the echo. However, if $\omega_{\beta,\alpha} = 2\pi n/\tau$,

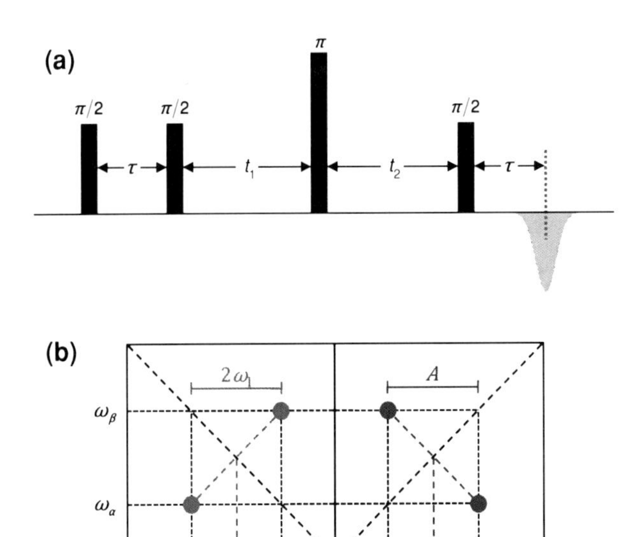

Fig. 2.19 **a** Pulse sequence for the hyperfine sublevel correlation (HYSCORE) experiment. **b** Schematic HYSCORE spectra. The spectral contribution of a weakly coupled nucleus manifests in the $(+,+)$ quadrant. Spectral features of a strongly coupled nucleus appear in the $(-,+)$ quadrant

a minimum is transferred to a minimum and a maximum is transferred to a maximum and the grating remains unaffected.

The distribution of blindspots in a HYSCORE spectrum with $\tau = 132$ ns is illustrated graphically in Fig. 2.20b. To assure the detection of all nuclear modulations, it is necessary to repeat experiments with different values of τ.

2.10.4 Phase Cycling

The $\pi/2 - \tau - \pi/2 - T - \pi/2$ pulse sequence does not only result in the formation of the stimulated echo, but also gives rise to unwanted echoes created by pairs of pulses (Fig. 2.21). The HYSCORE sequence generates even more unwanted echoes, which disturb the detected signal when evolution times are incremented.

These echoes can be efficiently suppressed by phase cycling. Here, one takes advantage of the fact that all echoes follow different coherence transfer pathways. With an appropriate phase cycle, only the pathway of the wanted echo is selected. Thus, the amplitude of this echo is maximized during the phase cycle, while all other echoes are annihilated.

A complete suppression of all echoes is achieved by a 4-step phase cycle for 3-pulse ESEEM and by a 16-step phase cycle for HYSCORE. However, since many coherence transfer paths do not lead to a refocusing, a 4-step phase

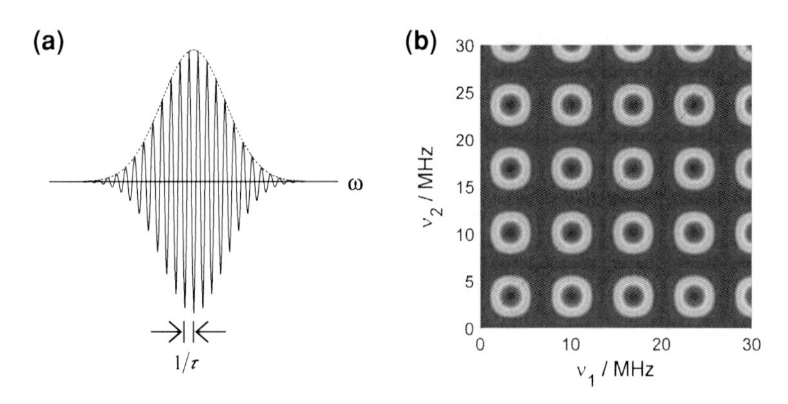

Fig. 2.20 **a** Polarization grating of a narrow EPR line created by a $\pi/2 - \tau - \pi/2$ pulse sequence. **b** Illustration of blind spots in the first quadrant of a HYSCORE spectrum for $\tau = 132$ ns assuming $|\omega_I| \gg |A/2|$. Spectral regions providing unsuppressed nuclear modulations are painted in warm colors while dark blue areas mark blindspots

cycle is usually sufficient [63]. In this thesis, an intermediate 8-step phase cycle was used.

2.11 Double Electron–Electron Resonance (DEER)

As introduced in Sects. 2.2.1 and 2.3.7, the characteristic dependence of dipole–dipole couplings can be utilized to retrieve distance distributions of two coupled electron spins. In the past, several pulse EPR methods were developed to separate the spin–spin interactions of electrons from other contributions of the spin Hamiltonian. Important single frequency techniques are the 2 + 1 method [64], SIFTER [65], and double quantum EPR [21, 66–68].

The application of two microwave frequencies (or two magnetic fields) allows for the spectral separation of a spin pair into an observer spin and a pump spin, which can be manipulated independently. In two-frequency methods, the separation of the spin–spin interactions is achieved by refocusing all interactions, including the spin–spin coupling by an echo sequence on the observer spin A. The spin–spin coupling is then reintroduced by a pump pulse on spin B that inverts the local magnetic field of spin A (Fig. 2.22c).

The selection of distinct observer and pump spins is also possible for a chemically identical spin pair. In this case, the anisotropy of the spectrum provides for the spectral separation. Typical observer and pump frequencies for nitroxides are shown in Fig. 2.22b. The frequency difference of ~65 MHz allows for non-overlapping excitation profiles.

The original 3-pulse electron–electron double resonance (ELDOR) method consists of a primary echo subsequence on A [69, 70]. The pump pulse on B is

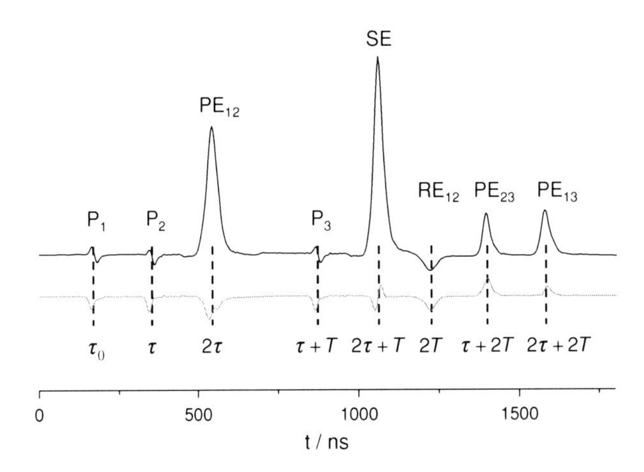

Fig. 2.21 Echoes generated by a $\pi/2 - \tau - \pi/2 - T - \pi/2$ pulse sequence applied to a 1:1 mixture of human serum albumin and 16-doxylstearic acid with $\tau = 180$ ns, $T = 600$ ns, and $t_P = 16$ ns. The real part is depicted in *black*, the imaginary part is given in *red*. The unwanted primary echoes (PE_{ij}) and the refocused echo, created by pairs of pulses (P_i,P_j), need to be annihilated by phase cyclings

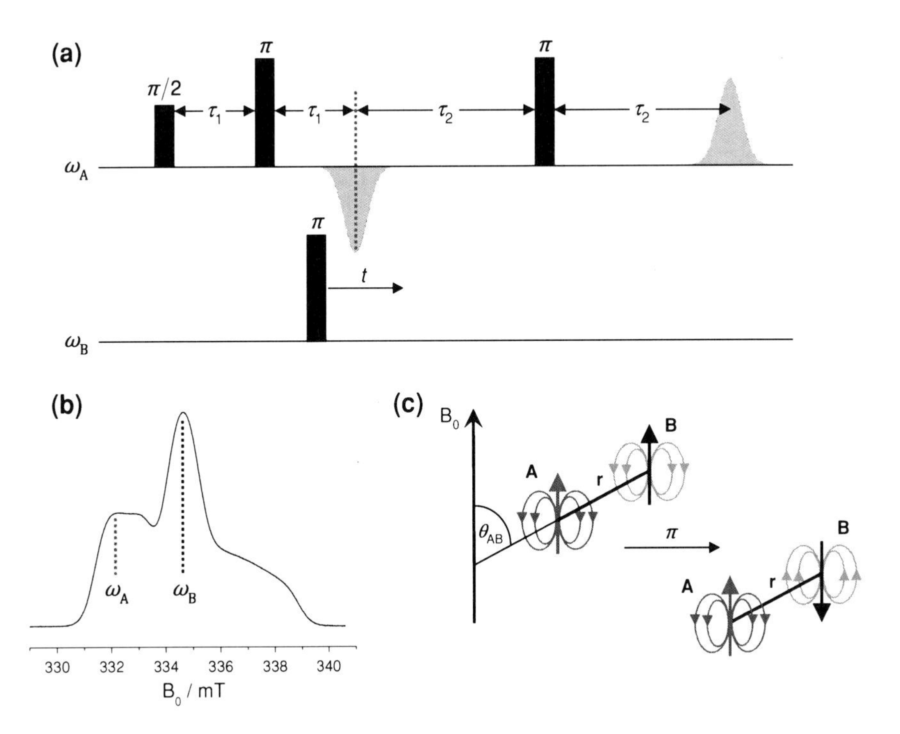

Fig. 2.22 a 4-Pulse sequence of the DEER experiment. **b** Typical nitroxide powder spectrum with spectral positions of the pump and observer frequencies. **c** Schematic illustration of the change of the local magnetic field caused by the pump π-pulse

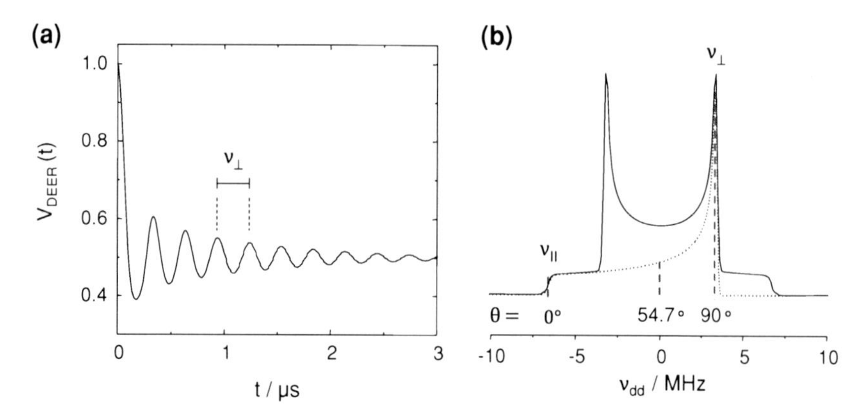

Fig. 2.23 **a** Calculated DEER time domain signal for two coupled nitroxides with $r = 2.5$ nm and $\sigma = 0.035$ nm and random orientational distribution in the matrix. **b** Corresponding Pake doublet in the frequency domain with characteristic frequency positions $\nu_\perp$ and $\nu_{||}$ and specification of the angular contributions

implemented between the first and the second pulse. The data acquisition of the dipolar evolution is however limited by the deadtime after the first pulse. This problem can be overcome by introduction of a second refocusing pulse on A. The pulse sequence of this 4-pulse double electron–electron resonance (DEER) technique is shown in Fig. 2.22a [19, 71]. By variation of the time delay between the second pulse of the observer sequence and the pump pulse, a dipolar modulation of the refocused echo is observed, given by

$$V_{\mathrm{DEER}}(r,t) = 1 - \int_0^{\pi/2} \sin\theta_{\mathrm{AB}}\lambda(\theta_{\mathrm{AB}})[1 - \cos(\omega_{\mathrm{DD}}(\theta_{\mathrm{AB}})t)]d\theta_{\mathrm{AB}} \tag{2.100}$$

with the dipolar frequency (cf. Eq. 2.42)

$$\omega_{\mathrm{DD}}(r,\theta_{\mathrm{AB}}) = \frac{\mu_0}{4\pi\hbar}\frac{g_{\mathrm{A}}g_{\mathrm{B}}\beta_{\mathrm{e}}^2}{r_{\mathrm{AB}}^3}(3\cos^2\theta_{\mathrm{AB}} - 1). \tag{2.101}$$

θ_{AB} denotes the angle between the static magnetic field and the connection vector of the coupled electron spins, and λ is the orientation-dependent modulation depth parameter. For nitroxides at X-band, no pronounced orientation selection is observed at the observer and pump frequencies and λ is assumed orientation-independent. In this case, a Pake doublet is obtained by Fourier transformation of the time-domain data (Fig. 2.23).

The lowest retrievable distance is limited by the excitation bandwidth of the pump pulse, which has to equal the width of the dipolar Pake pattern. A 12 ns pump pulse provides for a lower limit of 1.5 nm. Note that no pulse sequence provides for the separation of dipole–dipole coupling and spin exchange coupling.

However, the latter contribution is usually negligible for distances exceeding 1.5 nm [22]. The upper limit of the accessible distance range is given by the maximum recordable dipolar evolution time and thus by τ_2. To unambiguously retrieve a distance, at least 5/8 of the modulation needs to be recorded. For model biradicals in rigid matrices, $\tau_2 = 6\,\mu s$ and $r_{max} = 8\,nm$ are accessible. For biological systems in water, $\tau_2 = 2.5\,\mu s$ and $r_{max} = 6\,nm$ are more realistic values.

In analogy to the ESEEM experiments, the modulation is superimposed by an exponentially decaying signal,

$$V_{DEER}^{hom}(t) = \exp\left(-\frac{2\pi\mu_0 g_1 g_2 \beta_n}{9\sqrt{3}\hbar}\lambda c_B t^{d/3}\right). \tag{2.102}$$

This background function originates from a superposition of dipolar couplings to homogeneously distributed electron spins with distances up to 40 nm. It is dependent on the concentration of excited B spins λc_B and the fractal dimension d of the sample. Hence, the experimental time trace amounts to

$$V'_{DEER}(t) = V_{DEER}^{hom}(t)\, V_{DEER}(t). \tag{2.103}$$

For a narrow distance distribution, the singularities of the Pake doublet $\nu_\perp$ can be easily identified and the corresponding distance can be directly calculated. However, it is a non-trivial problem to retrieve broad or complicated distance distributions $G(r)$ from the time (or frequency) domain data. In the framework of shell factorization analysis, the spherical volume around the observer spin is subdivided into thin shells. The probability of finding a spin B in a shell with thickness dr at distance r from the observer spin A is $4\pi r^2 dr G(r)$, where $G(r)$ is the wanted distance distribution. The normalized DEER signal (without orientation selection) is then given by

$$\ln\frac{V_{DEER}(t)}{V_{DEER}(0)} = -4\pi\lambda \int_{r_{min}}^{r_{max}} r^2 G(r)(1 - V_r(r,t))dr, \tag{2.104}$$

where $V_r(r,t)$ denotes the time domain signal within a certain thin shell.

One way to solve this equation is to simulate the dipolar modulations with a model distance distribution that is varied to fit the experimental time trace. Another way is to calculate the distance distribution directly from the time domain data. This mathematical problem is ill-posed, i.e. a slight distortion of the experimental signal may lead to strong distortions in the distance distribution. During the last years, several routes have been proposed to improve the reliability of the results with Tikhonov regularization being the current method of choice [72]. It is implemented in the comprehensive software package Deer Analysis2008 by Jeschke, which is used to analyze the DEER data in Sect. 3.2 [73]. This method is, however, not applicable if λ shows a pronounced orientation dependence. In this case, the first route has to be chosen and the dipolar modulations need to be simulated (cf. Sect. 3.4).

References

1. Gerlach W, Stern O (1921) Z Phys 8:110–111
2. Gerlach W, Stern O (1922) Z Phys 9:349–352
3. Gerlach W, Stern O (1922) Z Phys 9:353–355
4. Zeeman P (1897) Philos Mag 43:226
5. Zeeman P (1897) Nature 55:347
6. Uhlenbeck GE, Goudsmit S (1925) Naturwissenschaften 13:953–954
7. Uhlenbeck GE, Goudsmit S (1926) Nature 117:264–265
8. Zavoisky E (1945) J Phys USSR 9:211
9. Zavoisky E (1945) J Phys USSR 9:245
10. Purcell EM, Torrey HC, Pound RV (1946) Phys Rev 69:37–38
11. Bloch F, Hansen WW, Packard M (1946) Phys Rev 69:127
12. Ernst RR, Anderson WA (1966) Rev Sci Instrum 37:93–102
13. Blume RJ (1958) Phys Rev 109:1867–1873
14. Holczer K, Schmalbein D (1987) Bruker Rep 1:22
15. Höfer P, Maresch GG, Schmalbein D, Holczer K (1996) Bruker Rep 142:15
16. Likhtenshtein GI, Yamauchi J, Nakatsuji S, Smirnov AI, Tamura R (2008) Nitroxides—applications in chemistry, biomedicine and materials science. Wiley-VCH, Weinheim
17. Jeschke G (2006) Site-specific information on macromolecular materials by combining CW and pulsed ESR on spin probes. In: Schlick S (ed) Advanced ESR methods in polymer research. Wiley, New York
18. Hinderberger D, Jeschke G (2006) Site-specific characterization of structure and dynamics of complex materials by EPR spin probes. In: Webb GA (ed) Modern magnetic resonance. Springer, London
19. Jeschke G, Pannier M, Spiess HW (2000) Double electron–electron resonance. In: Berliner LJ, Eaton GR, Eaton SS (eds) Biological magnetic resonance, vol 19: distance measurements in biological systems by EPR. Kluwer Academic, New York
20. Jeschke G (2002) Chem Phys Chem 3:927–932
21. Borbat PP, Freed JH (2000) Double-quantum ESR and distance measurements. In: Berliner LJ, Eaton GR, Eaton SS (eds) Biological magnetic resonance, vol 19: distance measurements in biological systems by EPR. Kluwer Academic, New York
22. Jeschke G (2002) Macromol Rapid Commun 23:227–246
23. Schiemann O, Prisner TF (2007) Q Rev Biophys 40:1–53
24. Borbat PP, Costa-Filho AJ, Earle KA, Moscicki JK, Freed JH (2001) Science 291:266–269
25. Jeschke G, Polyhach Y (2007) Phys Chem Chem Phys 9:1895–1910
26. Schmidt-Rohr K, Spiess HW (1994) Multidimensional solid-state NMR and polymers. Academic Press, London
27. Dockter C, Volkov A, Bauer C, Polyhach Y, Joly-Lopez Z, Jeschke G, Paulsen H (2009) Proc Nat Acad Sci U S A 106:18485–18490
28. Hinderberger D, Spiess HW, Jeschke G (2010) Appl Magn Reson 37:657–683
29. Odom B, Hanneke D, D'Urso B, Gabrielse G (2006) Phys Rev Lett 97:030801
30. Bloch F (1946) Phys Rev 70:460–474
31. Abragam A, Pryce MHL (1951) Proc R Soc A: Math Phys Sci 205:135–153
32. Solomon EI, Hodgson KO (1998) Spectroscopic methods in bioinorganic chemistry. Clarendon, Oxford
33. McGarvey BR (1967) J Phys Chem 71:51–66
34. Molin YN, Salikhov KM, Zamaraev KI (1980) Spin exchange–principles and applications in chemistry and biology. Springer Verlag, Berlin
35. McConnell HM (1958) J Chem Phys 28:430–431
36. Hubbell WL, Cafiso DS, Altenbach C (2000) Nat Struct Biol 7:735–739
37. Kocherginsky N, Swartz HM (1995) Nitroxide spin labels - reactions in biology and chemistry. CRC Press, Boca Raton

38. Fajer PG, Brown L, Song L (2007) Practical pulsed dipolar ESR (DEER). In: Hemminga MA, Berliner LJ (eds) Biological magnetic resonance, vol 27: ESR spectroscopy in membrane biophysics. Springer, New York
39. D. Hinderberger (2004) Polyelectrolytes and their counterions studied by EPR spectroscopy, Doctoral Dissertation, Johannes Gutenberg-Universität, Mainz,
40. Beth AH, Robinson BH (1989) Nitrogen-15 and deuterium substituted spin labels for studies of very slow rotational motion. In: Berliner LJ, Reuben J (eds) Biological magnetic resonance, vol 8: Spin labeling—theory and applications. Plenum Press, New York
41. Edmonds AR (1960) Angular momentum in quantum mechanics. Princeton University Press, Princeton
42. Jeschke G, Schlick S (2006) Continuous-wave and pulsed ESR methods. In: Schlick S (ed) Advanced ESR methods in polymer research. Wiley, Hoboken
43. Kivelson D (1960) J Chem Phys 33:1094–1106
44. Freed JH, Fraenkel GK (1963) J Chem Phys 39:326–348
45. Schneider DJ, Freed JH (1989) Continuous-wave and pulsed ESR methods. In: Berliner LJ, Reuben J (eds) Biological magnetic resonance, vol 8: spin labeling–theory and applications. Plenum Press, New York
46. Meirovitch E, Freed JH (1984) J Phys Chem 88:4995–5004
47. Polimeno A, Freed JH (1995) J Phys Chem 99:10995–11006
48. Wegener C, Savitsky A, Pfeiffer M, Möbius K, Steinhoff HJ (2001) Appl Magn Reson 21:441–452
49. Bordignon E, Steinhoff HJ (2007) Membrane protein structure and dynamics studied by site-directed spin-labeling ESR. In: Hemminga MA, Berliner LJ (eds) Biological magnetic resonance, vol 27: ESR spectroscopy in membrane biophysics. Springer, New York
50. Stoll S, Schweiger A (2006) J Magn Reson 178:42–55
51. Stoll S, Schweiger A (2007) Easy spin: simulating cw ESR spectra. In: Hemminga MA, Berliner LJ (eds) Biological magnetic resonance, vol. 27: ESR spectroscopy in membrane biophysics. Springer, New York
52. Weil JA (1971) J Magn Reson 4:394–399
53. Goldman SA, Bruno GV, Polnaszek CF, Freed JH (1972) J Chem Phys 56:716–735
54. Hwang JS, Mason RP, Hwang LP, Freed JH (1975) J Phys Chem 79:489–511
55. Stoll S, Schweiger A (2003) Chem Phys Lett 380:464–470
56. Sørensen OW, Eich GW, Levitt MH, Bodenhausen G, Ernst RR (1983) Prog Nucl Magn Reson Spectrosc 16:163–192
57. Sørensen OW (1989) Prog Nucl Magn Reson Spectrosc 21:503–569
58. Hahn EL (1950) Phys Rev 80:580–594
59. Rowan LG, Hahn EL, Mims WB (1965) Phys Rev 137:A61–A71
60. Mims WB (1972) Phys Rev B 5:2409–2419
61. Mims WB, Davis JL, Peisach J (1990) J Magn Reson 86:273–292
62. Höfer P, Grupp A, Nebenführ H, Mehring M (1986) Chem Phys Lett 132:279–282
63. Gemperle C, Aebli G, Schweiger A, Ernst RR (1990) J Magn Reson 88:241–256
64. Kurshev VV, Raitsimring AM, Tsvetkov YD (1989) J Magn Reson 81:441–454
65. Jeschke G, Pannier M, Godt A, Spiess HW (2000) Chem Phys Lett 331:243–252
66. Saxena S, Freed JH (1996) Chem Phys Lett 251:102–110
67. Saxena S, Freed JH (1997) J Chem Phys 107:1317–1340
68. Borbat PP, Freed JH (1999) Chem Phys Lett 313:145–154
69. Milov AD, Salikhov KM, Shirov MD (1981) Fiz Tverd Tela 23:975–982
70. Milov AD, Maryasov AG, Tsvetkov YD (1998) Appl Magn Reson 15:107–143
71. Pannier M, Veit S, Godt A, Jeschke G, Spiess HW (2000) J Magn Reson 142:331–340
72. Jeschke G, Panek G, Godt A, Bender A, Paulsen H (2004) Appl Magn Reson 26:223–244
73. Jeschke G, Chechik V, Ionita P, Godt A, Zimmermann H, Banham J, Timmel CR, Hilger D, Jung H (2006) Appl Magn Reson 30:473–498

Bibliography

74. Atherton NM (1973) Principles of electron spin resonance. Ellis Horwood, Chichester
75. Weil A, Bolton JR, Wertz JE (1994) Electron paramagnetic resonance: elementary theory and practical applications. Wiley, New York
76. Schweiger A, Jeschke G (2001) Principles of pulse electron paramagnetic resonance. Oxford University Press, Oxford
77. Hore PJ, Jones JA, Wimperis S (2000) NMR: The toolkit. Oxford University Press, Oxford
78. Jeschke G (1998) Einführung in die ESR-Spektroskopie (lecture notes), Mainz
79. Jeschke G (2008) Kurze Einführung in die elektronenparamagnetische Resonanzspektroskopie (lecture notes), Konstanz

Chapter 3
The Functional Structure of Human Serum Albumin

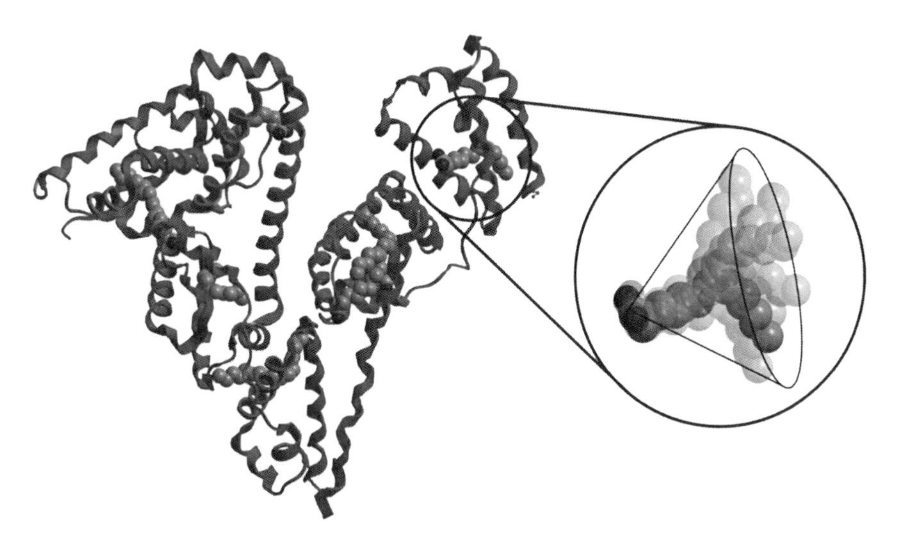

Human serum albumin (HSA) is a versatile transport protein for various endogenous compounds and drugs. This study focuses on its highly relevant transport function for fatty acids in the circulatory system. While extensive crystallographic data on HSA–fatty acid binding exist, a new spectroscopic approach is used to gain information on the functional structure of HSA in solution. Using spin-labeled stearic acid and applying double electron–electron resonance (DEER) spectroscopy, the functional protein structure is accessed for the first time from the ligands' point of view. Spatial distributions of the anchoring groups and entry points of the fatty acid binding sites are obtained studying fatty acids with different labeling position. While the distribution of the head groups is mainly consistent with the crystallographic data, the entry points of the binding sites are distributed much more homogeneously on the protein surface than suggested by the crystal

M. J. N. Junk, *Assessing the Functional Structure of Molecular Transporters by EPR Spectroscopy*, Springer Theses, DOI: 10.1007/978-3-642-25135-1_3,

structure. This symmetric distribution provides a straightforward explanation for the transport function of the protein as it facilitates a fast uptake and release of multiple fatty acids.

3.1 Introduction

Human serum albumin (HSA) is the most abundant protein in human blood plasma and serves as a transporting agent for various endogenous compounds and drug molecules [1, 2]. Especially, its capability to bind and transport multiple fatty acids (FA) has been studied extensively in the past [3, 4]. The research on HSA was severely hampered by the complexity of the protein and benefitted a lot from crystallographic high-resolution structures. Nearly 20 years ago, He and Carter reported the first crystal structure [5]. Up to date, a plentitude of crystal structures has been deposited in the Protein Data Bank.

Due to the pioneering work of Curry et al., crystal structures of various HSA–fatty acid complexes have become accessible allowing new insights into the FA binding properties of the protein [6–8]. In particular, they found that fatty acids are distributed highly asymmetrically in the protein crystal despite the fact that HSA exhibits a symmetric primary and secondary structure. Up to seven distinct binding sites were found for long chain fatty acids, most of which comprised of ionic anchoring units and long, hydrophobic pockets [8, 9]. The location of two to three high affinity binding sites [3, 10] was assigned by correlation of the X-ray structure with NMR studies on competitive binding of drugs replacing ^{13}C-labeled fatty acids [11, 12]. Sites 2, 4, and 5 bind fatty acids with a high affinity, while sites 1, 3, 6, and 7 exhibit a somewhat lower affinity to fatty acids (see Fig. 3.1a).

On a more general note, there is a long standing debate to what extent protein crystal structures reflect the dynamic and functional structures of proteins in solution. This debate is often fueled by apparent discrepancies between X-ray crystallographic data and results from solution-state based techniques (e.g. NMR and other types of spectroscopy as well as neutron scattering) or from molecular dynamics simulations. Moreover, there is an increasing awareness that protein dynamics in solution is connected to its biological function. Recent NMR studies revealed that many proteins exhibit pronounced dynamic conformational flexibilities [13–15].

It is well known that, in particular, surface exposed parts of HSA show a high degree of flexibility which constitutes a key to the protein's binding versatility towards various molecules. Already in the 1950s, Karush developed a concept which accounted for this conformational adaptability of the binding sites [16, 17]. Further, a model has been proposed, which takes into account the conformational entropy arising from the flexibility of the fatty acid alkyl chains [18].

This study aims at unraveling the functional structure of HSA with respect to its binding of fatty acids. An electron paramagnetic resonance (EPR) technique is applied to study the fatty acid binding site distribution in the protein in frozen solution by characterizing it from the fatty acids' point of view. This is achieved

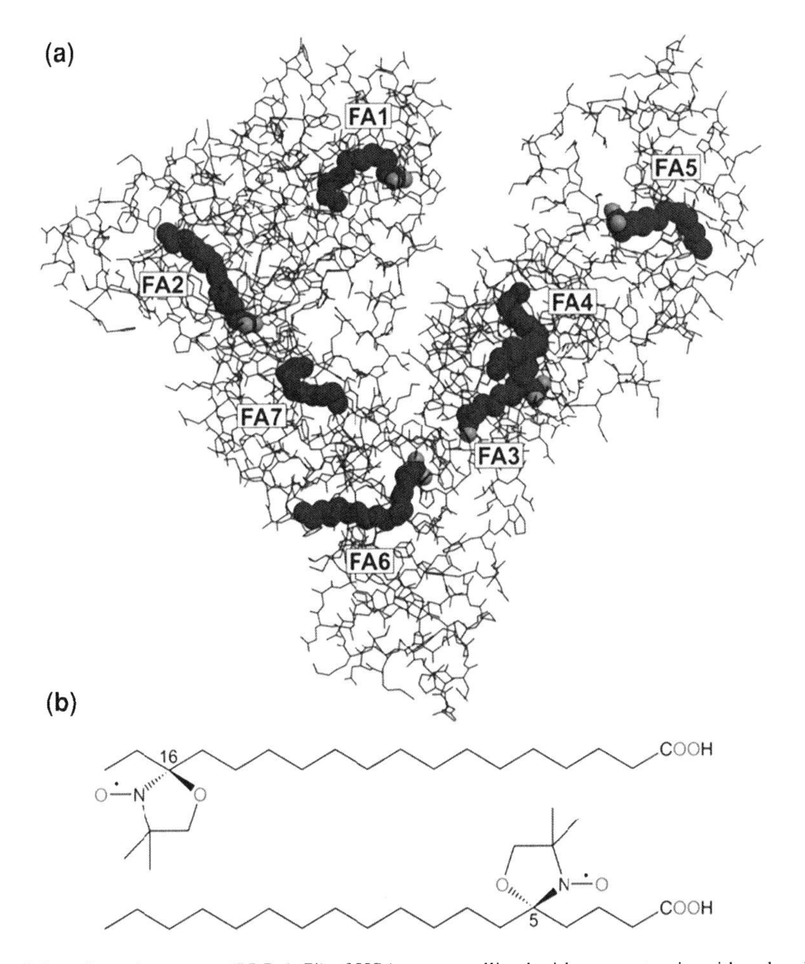

Fig. 3.1 **a** Crystal structure (PDB 1e7i) of HSA co-crystallized with seven stearic acid molecules [8]. The oxygen atoms of the FA carboxylic acid head groups are displayed in *red*. **b** Chemical structure of the EPR active molecules, 5-doxylstearic acid (DSA) and 16-DSA. Reprinted with permission from [65, 66]. Copyright 2010 WILEY-VCH Verlag GmbH and Co. KGaA, Weinheim

by studying spin-labeled fatty acids, which alone give rise to an EPR signal [19–21]. Thus, the distribution of the FA binding sites is detected without any contribution from the complex protein itself (Fig. 3.1a). Structural information of the binding sites is obtained by determining the distance distributions between the fatty acids. These distance distributions are retrieved by double electron–electron resonance (DEER), a pulse EPR method, which utilizes the inherent distance dependence of the dipolar couplings (acting solely through space) between the unpaired electron spins [22–24]. In recent years, DEER has increasingly been used in structural studies on both synthetic [25, 26] and biological systems [27] with the focus on (membrane) proteins and nucleic acids [28–31].

To sample distance information from different positions along the methylene chain of the fatty acids in the respective binding sites, fatty acids with different labeling positions were applied. In 5-doxylstearic acid (5-DSA), the unpaired electron resides near the anchoring carboxylic acid group. In 16-DSA, it is located near the end of the methylene chain (Fig. 3.1b). Thus, information can be retrieved from the anchor positions in the protein as well as from the entry points into the fatty acid channel formed by the protein.

This chapter is divided into three parts. In the first section, the uptake of spin-labeled fatty acids by the protein is characterized with CW EPR spectroscopy. Following this, distance distributions are recorded when loading the protein with various amounts of fatty acids. The experimental results are then compared with the expected distributions as derived from the crystal structure and differences are discussed in terms of the protein function. In a more technical second part, spin systems with more than two unpaired electrons per protein molecules are studied. Spin counting is applied to quantify the number of coupled spins and limitations of the method for self-assembled systems are discussed. Further, distortions in the distance distribution due to multispin effects are characterized in detail. In the third section, a second compound is admixed to the protein beside fatty acids. Cu(II) protoporphyrin IX serves as an EPR probe with a large g anisotropy. Via orientationally selective DEER, not only the distances but also the relative orientations of the fatty acid in the protein are accessed and compared to theoretical data as suggested by the crystal structure.

3.2 The Distribution of Fatty Acids in Human Serum Albumin in Solution

3.2.1 Results

All experiments in this chapter imply the addition of up to seven fatty acids to one HSA molecule to occupy all binding sites in the protein. To avoid artifacts and allow quantitative interpretation of the data, it is essential to limit the amount of EPR active, spin-labeled fatty acids to two per protein molecule. This avoids complications due to multispin effects (cf. Sect. 3.3) [32]. By simultaneously adding diamagnetic fatty acids, the degree of loading can be still varied in such *spin-diluted* systems, as the diamagnetic fatty acids compete with DSA for a binding site without giving rise to an EPR signal. By adjusting the ratio of diamagnetic fatty acid and DSA, two sites, statistically distributed among all sites, are then occupied by EPR active molecules. Thus, an artifact-free distance distribution with a complete set of distances from all fatty acid binding sites can be obtained.

The diamagnetic fatty acid was prepared by reduction of the corresponding DSA to the EPR-inactive hydroxylamine (rDSA). Details of the reaction are given in Appendix A.1.1. This molecule is structurally closely related to the paramagnetic DSA and exhibits comparable binding affinities, as checked by CW EPR

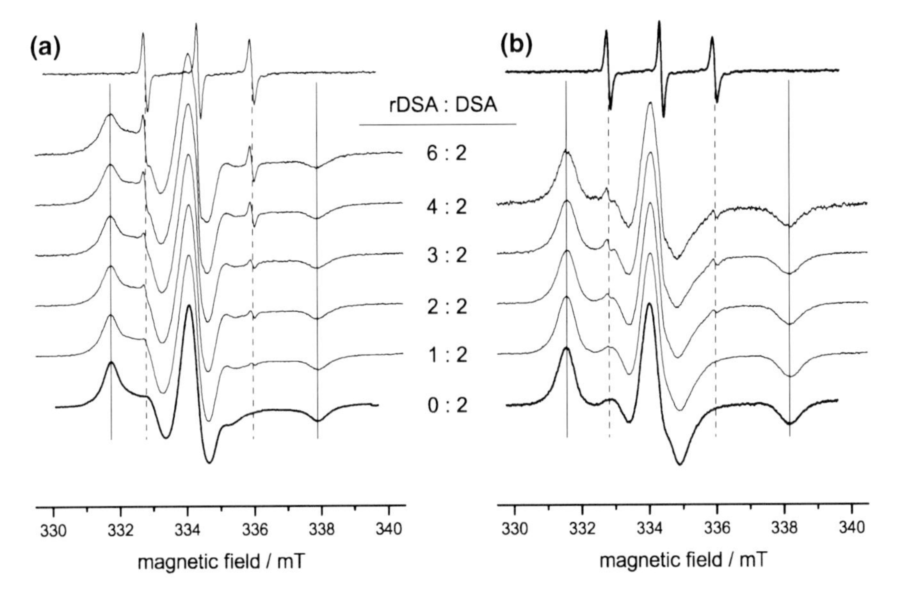

Fig. 3.2 CW EPR spectra of HSA–fatty acid mixtures with **a** 16-DSA and **b** 5-DSA recorded at 298 K. The rDSA:DSA values indicate the average number of reduced and EPR-active fatty acid per protein molecule. The characteristic signatures of the respective DSA bound to HSA are marked by solid lines, the signature of free fatty acids in solution is marked by dashed lines. For comparison, a reference spectrum of 0.2 mM DSA in aqueous solution is displayed on top. Reprinted with permission from [65, 66]. Copyright 2010 WILEY-VCH Verlag GmbH and Co. KGaA, Weinheim

spectroscopy (Appendix A.1.2). Stearic acid could also be used as spin diluting paramagnetic species, but suffers from a low solubility and a larger structural variation in comparison to DSA. Moreover, similar results are obtained independent of the nature of the spin diluting stearic acid (Appendix A.1.1).

Uptake of spin labeled fatty acids

The CW EPR spectra of both types of spin-labeled fatty acids, 5-DSA and 16-DSA, at different HSA–fatty acid ratios are displayed in Fig. 3.2. The spectra diplay signatures of rotationally restricted and freely tumbling nitroxides, as indicated by solid and dashed lines. These two species correspond to fatty acid molecules bound to DSA and free fatty acids in solution.

Up to a HSA–fatty acid ratio of 1:3, no (5-DSA) or only a negligible (16-DSA) signature of free species is observed, indicating complete uptake of the fatty acids by the protein. This confirms the existence of three high affinity binding sites in full agreement with the literature. At higher fatty acid ratios, the relative amount of unbound DSA steadily increases. Yet, at a HSA–fatty acid ratio of 1:6, more than 99.7% of all fatty acids are still complexed by the protein as deduced from the spectral intensities of both species. This proves a nearly quantitative uptake of both types of DSA and confirms that the fatty acid uptake is not disturbed by spin labeling.

Several publications dealing with the fatty acid binding properties of HSA use commercial HSA that is labeled fatty acid free. CW EPR studies, however, revealed a decreased uptake of fatty acids in such samples, which is indicative of partial protein degeneration (Appendix A.1.3). Hence, these studies were performed with commercially available HSA that was not explicitly fatty acid free but labeled non-denaturated. Potentially residual fatty acid molecules were regarded as a far less severe problem, since they can exchange with the EPR-active fatty acids and do not hamper a desired uniform distribution of DSA among all possible binding sites.

The degree of rotational freedom of the nitroxide can be estimated from the distance between the low field maximum and the high field minimum by spectral simulation [33, 34]. 5-DSA is rotationally more restricted than 16-DSA which has a roughly three times lower rotational correlation time. This trend is expected since the nitroxide unit of 5-DSA is placed in the tight hydrophobic binding channels near the anchoring point. The nitroxide group of 16-DSA, on the other hand, is located at the end of such a channel or even exceeds the end of a channel and will thus experience a higher degree of rotational freedom as reported by detailed CW EPR studies on the binding of spin-labeled fatty acids to different types of serum albumin [19–21].

Distances between the acid sites
To retrieve the desired information about the functional structure of HSA, the HSA–fatty acid complexes were further analyzed by DEER spectroscopy. The intramolecular part of the time-domain data and the extracted distance distributions are displayed in Fig. 3.3. For 16-DSA, well-defined dipolar modulations are observed which originate from narrow distance distributions with a dominating distance at 3.6 nm and two smaller contributions at 2.2 and 4.9 nm. For 5-DSA, such a pronounced modulation is missing, which is indicative of a broader distribution. In fact, a broad single distance peak is derived, which covers a range from 1.5 nm (the lower accessible limit with DEER) to 4 nm with a maximum around 2.5 nm.

Remarkably, neither the 5-DSA nor the 16-DSA distance distributions change considerably when the protein is loaded with different amounts of fatty acids. Thus, higher loading does not result in the generation of new distinct distance contributions. It rather results in a broadening of the already existing distance peaks. This can be explained as follows. Although fatty acids possess different binding affinities to different protein binding sites, these differences seem not very pronounced. Hence, the binding sites are not filled up consecutively. Rather, all binding sites are populated to a certain degree even at low HSA–fatty acid ratios. The high affinity binding sites are only populated to a larger extent. Variations of the HSA–fatty acid ratio will then only lead to changes of the relative populations. This explanation is supported by NMR studies with ^{13}C labeled fatty acids. Only subtle changes of the NMR peak ratios were observed when increasing the fatty acid ratio [11, 12].

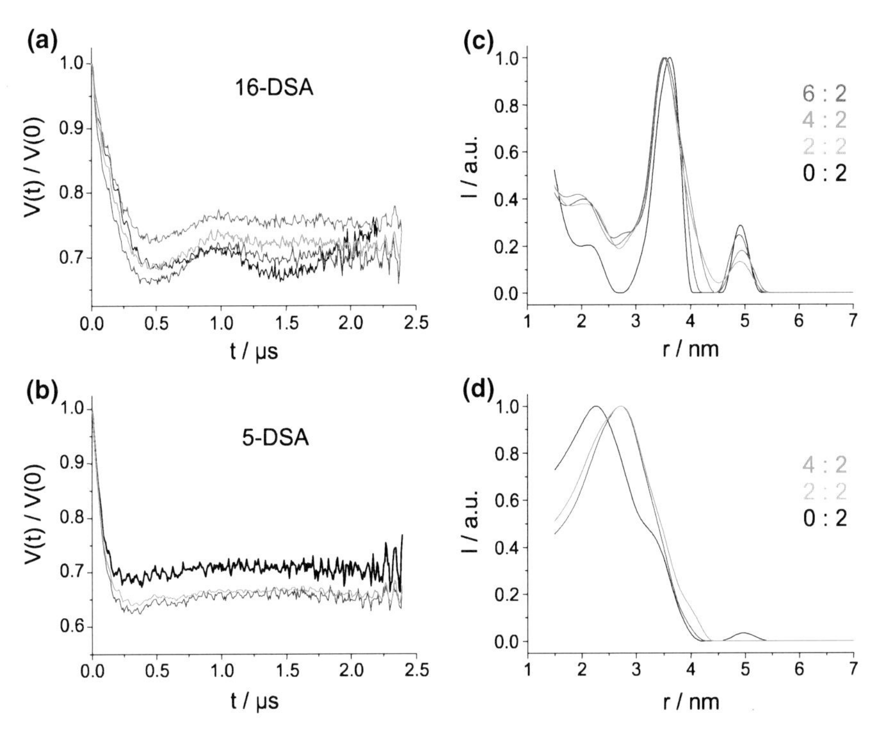

Fig. 3.3 **a, b** Intramolecular part of the DEER time-domain data and **c, d** extracted distance distributions of spin-labeled stearic acids complexed in HSA with varying numbers of reduced and EPR active fatty acid per protein molecule: *black* (0:2), *blue* (2:2), *green* (4:2), *orange* (6:2). The data from 16-DSA are shown on *top* (**a, c**), the data from 5-DSA on the bottom (**b, d**). Reprinted with permission from [65, 66]. Copyright 2010 WILEY-VCH Verlag GmbH & Co. KGaA, Weinheim

3.2.2 Discussion

While the CW EPR data were collected at 298 K, DEER measurements were conducted at 50 K. For this purpose, the solution containing the HSA–fatty acid complexes was shock-frozen to obtain a vitrified solution. Thus, a snapshot of the protein ensemble in solution at the glass transition temperature of around 170 K is studied [35, 36]. Note that 20 vol% of glycerol were added to obtain such glassy samples. This is a standard treatment in pulse EPR and there is plenty of evidence that this small amount of non-aqueous solvent does not alter the structure of proteins but rather stabilizes the native structure in frozen solutions [28, 37]. This assumption is supported by CW EPR measurements at 298 K that did not show changes of the fatty acid binding dependent on the addition of glycerol. In contrast, only small amounts of ethanol (10 vol%) are sufficient to denature the tertiary structure of the protein. In this case, the DEER time-domain signal consists of a homogeneous exponential decay without underlying modulations (data not shown).

In this study, spin-labeled DSA is used as a substitute for natural fatty acids. The doxyl EPR reporter group causes only a minor structural variation, but could lead to different binding affinities since some fatty acid binding sites are reported as narrow tunnels. In this case, the presence or absence of a bulky doxyl group could be relevant. For 16-DSA, the spin label is located at the end of the methylene tail and, at least for some sites, is probably not even required to enter the binding site. In contrast, the doxyl group of 5-DSA is located in the vicinity to the ionic head group of the fatty acid. It may thus be argued that this alteration of the FA structure may result in potentially less efficient binding to HSA. However, CW EPR data reveal that both spin-labeled fatty acids are taken up equally by the protein. Both DSA variants are complexed almost completely at a HSA–fatty acid ratio of 1:6. Furthermore, they display competitive binding affinities in comparison to stearic acid. Thus, neither steric nor electronic variations severely affect the binding properties of the spin-labeled fatty acid. Considering the wide variety of saturated and unsaturated FA that can be bound by HSA [3, 8], it is thus not surprising that the binding of DSA molecules is comparable to that of other long chain fatty acids.

The different labeling positions allow for two different views on the functional structure of the protein. The position of the unpaired electron in 5-DSA is close to the carboxylic acid group of the fatty acid, which interacts with positively charged side groups of the protein. Hence, DEER delivers the characterization of the spatial distribution of the anchoring groups. With 16-DSA, on the other hand, the entry points into the fatty acid channels are probed. Information about the spatial distribution of these points is important to gain a better understanding of the uptake and release properties of the protein.

The 5-DSA (and thus the headgroup) distribution in solution is much broader than the distribution of 16-DSA (i.e. the entry points). Despite the increased flexibility of the doxyl moiety of 16-DSA, a very uniform distribution with a well-defined main distance is obtained. This suggests that the entry points of the fatty acid binding sites are distributed rather symmetrically on the surface of the protein molecule.

In Fig. 3.4, the experimental distance distributions in solution are compared with distributions retrieved from the crystal structure. These distributions were calculated assuming a full occupation of all binding sites and a Gaussian broadening of the distance peaks. The procedure is detailed in the experimental part (Sect. 3.6) and in Appendix A.1.6. Since the fatty acids in sites 1 and 7 are not resolved up to the C-16 atom of the methylene chain, they were extrapolated to this position. Additionally, Fig. 3.4 contains an alternative distribution without these two low affinity binding sites.

The experimental distribution of 5-DSA show major similarities to that of the crystal structure assuming a full occupation of all binding sites. In both cases, broad distributions centered at a distance of about 3 nm are obtained. Indeed, the highly asymmetric distribution of fatty acid binding sites observed by crystallographic data can also be found in the DEER results, namely for position C-5, i.e. for the anchoring points. In that case, the distributions from both methods coincide

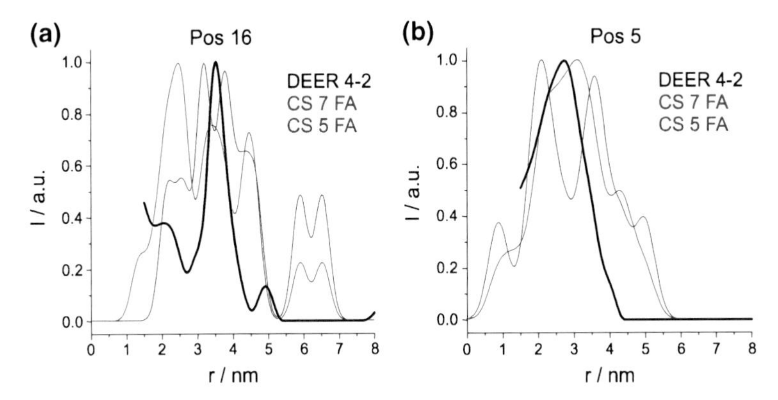

Fig. 3.4 Comparison of experimental distance distributions obtained by DEER (*black*) with calculated distributions obtained from the crystal structure (*red, blue*) for the **a** C-16 position and **b** C-5 position. The distribution in red is obtained assuming that all seven binding sites occupied. The distribution in *blue* results when the fatty acids in low affinity binding sites 1 and 7 are neglected which are not completely resolved in the crystal structure. Note that distances >6 nm cannot be accessed by DEER under the applied conditions. Reprinted with permission from [65, 66]. Copyright 2010 WILEY-VCH Verlag GmbH and Co. KGaA, Weinheim

remarkably well. First of all, this shows that the distance distributions determined from DEER are reliable. Second, it suggests that the anchoring point distribution in solution reflects the more rigid inner part of the protein, which does not differ too strongly in the crystal and in solution.

In contrast, the distance distribution of the entry points (16-DSA) strongly deviates from the crystal structure, taking into account that distances greater than 6 nm cannot be accessed by DEER under the applied conditions. The crystal structure distribution exhibits three major peaks at 2.5, 3.5, and 4.5 nm. While the peak positions roughly agree with the DEER data, the relative intensities deviate considerably. Experimentally, the peak around 3.6 nm by far dominates all other peaks, which remarkably simplifies the distance distribution as compared to that derived from the crystal structure. It suggests that the entry points are distributed much more symmetrically and homogeneously over the protein surface than it would be expected from the crystal structure. Note that it is even impossible to reconcile the dominant (5- and 16-DSA) distances found from DEER measurements with those of the crystal structure. This is explained in detail in Appendix A.1.7. A homogeneous distance distribution of six binding sites suggests high symmetry. Considering the six sites to form an octahedron, on expects 12 vertex–vertex distances with length r and 3 distances with length $\sqrt{2}r$. With the dominating distance $r = 3.6$ nm, a diagonal distance of 5.1 nm results, which is in remarkable agreement with the observed distance of 4.9 nm. Indeed, an octahedron constitutes the most favorable distribution for the entries of six fatty acid binding sites on a sphere, as they are then easily accessible from every side of the protein and assume the maximum distance with respect to each other. However, such an octahedron model can only account for the entry points of six fatty acid binding sites. Moreover, it does not account for the structure of the protein, which is rather heart-shaped than sphere-like.

One may speculate that the more homogeneous distribution of the entry points arises from the conformational flexibility of HSA. Furthermore, it may even mirror the optimization of the protein to allow for a fast and facilitated uptake and release of the fatty acids. For this, an optimized average distribution of FA entry points as well as large conformational flexibility are prerequisites. A large flexibility on or close to the protein surface may also be entropically favored. The gain in entropy by only small conformational variations is much larger than for changes in the interior of the protein.

A homogeneous distribution of binding sites on an octahedron is also in accord with the fact that the DEER distance distributions effectively do not change when the fatty acid ratio is varied. Even if the binding sites were filled up consecutively, they would all give rise to similar distances due to their symmetric distribution on the protein surface.

To check the probability of such a symmetric distribution of the entry points, a molecular dynamics (MD) simulation on the crystal structure (1e7i) was performed for a period of 6.5 ns. Though no substantial deviation from the overall crystal structure is observed, it was found that the C-16 position of the fatty acids is more strongly affected by small changes of secondary structure elements than position C-5 (for details, see Appendix A.1.8). This is an additional hint that the anchoring region of the protein is rigid while the entry points have a substantial higher degree of freedom and flexibility.

3.3 Multispin Contributions to DEER Spectra

In this section, it will be studied in detail how the DEER data are affected when multiple paramagnetic centers are clustered in the protein. Special emphasis is placed on the differences between these multispin systems and the spin-diluted systems that were discussed in the previous section. Specifically, it is elucidated here how multispin interactions affect the experimental distance distributions and hamper the correct interpretation of the data.

3.3.1 Spin Counting

In Fig. 3.5, the intramolecular parts of the DEER time-domain data and the corresponding distance distributions are shown when the protein is loaded with 1–8 paramagnetic 16-DSA molecules. At first sight, the distance distributions for high loading exhibit considerable deviations to the distributions of the spin-diluted systems displayed in Fig. 3.3b. This issue will be discussed in detail in the next part of this section, while in the subsequent paragraphs the DEER time-domain data of HSA complexed with multiple EPR-active 16-DSA units is analyzed.

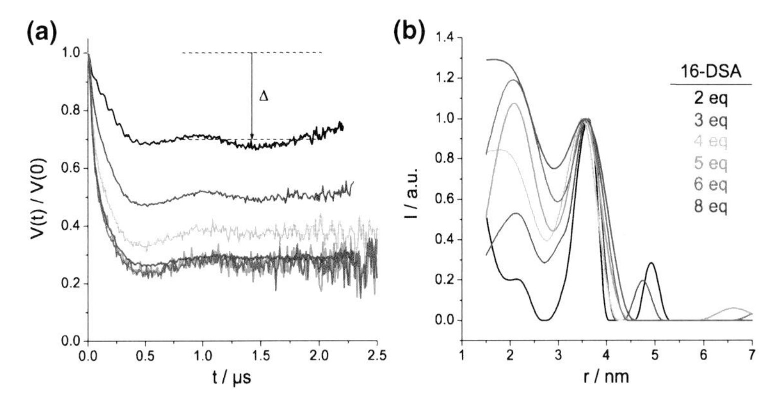

Fig. 3.5 **a** Background corrected DEER time-domain data of HSA complexed with 2–8 equivalents of 16-DSA with a graphical illustration of the definition of the modulation depth Δ. **b** Corresponding distance distributions by Tikhonov regularization with a regularization parameter of 100 normalized to the height of the peak at 3.6 nm. Reprinted with permission from [67]. Copyright 2011 Elsevier Inc

As mentioned in the last section, the fraction of fatty acids bound to HSA can be determined by CW EPR measurements. It was already discussed that 16-DSA is almost completely bound to the protein in HSA–fatty acid mixtures up to a ratio of 1:6 (cf. Appendix A.1.2). DEER provides a different means to quantify the fatty acid molecules per protein by the determination of the number of dipolar coupled spins N [38], which can be accessed by the relation [39]

$$N = \frac{\ln(1-\Delta)}{\ln(1-\lambda)} + 1. \tag{3.1}$$

Δ is the modulation depth and λ the inversion efficiency. The modulation depth, illustrated graphically in Fig. 3.5a is defined by

$$\Delta = 1 - \lim_{t\to\infty} V(t)/V(0) \tag{3.2}$$

and can easily be determined from the background corrected DEER time trace. The inversion efficiency is accessed experimentally using a biradical ($N = 2$, $\Delta = \lambda$). Note that λ is proportional but not identical to the experimental inversion efficiency λ_{exp} that is mentioned later in this section. The latter variable characterizes the inversion efficiency of selected spin packets in a narrow frequency range, while λ quantifies the average inversion efficiency for the pumped spins.

As can be seen in Fig. 3.5a, the modulation depth increases as more spin-labeled fatty acid equivalents are added to the protein. It reaches a constant value for HSA–fatty acid ratios $\geq$ 1:5. As expected, the increase of the average number of spins per protein molecule is reflected in the increase of the modulation depth.

The inversion efficiency was determined with a rigid phenylene-ethynylene based biradical with a spin–spin distance of 2.8 nm dissolved in *o*-terphenyl [40]. λ depends on a variety of parameters such as the length of the pump pulse, the

Table 3.1 Modulation depths and average numbers of spins in HSA–fatty acid Complexes (reproduced with permission from [67])

	Biradical	HSA : 16-DSA					
		1:2	1:3	1:4	1:5	1:6	1:8
Δ	0.516	0.300	0.497	0.626	0.728	0.730	0.708
N (Ref: biradical)	2.0	1.49	1.94	2.36	2.80	2.81	2.70
N (Ref: 1:2)	–	2.0	2.93	3.76	4.65	4.67	4.45

shape and width of the resonator mode, the spectral line shape, and the position of the pump pulse in the spectrum and in the resonator mode [41]. For this reason, all experimental conditions were kept constant for the biradical and the HSA–fatty acid complexes.

However, using the biradical as a reference, the average number of coupled spins varies from 1.49 to 2.81 for all HSA–DSA mixtures (Table 3.1). This suggests that the protein is able to complex less than three DSA molecules on average, which is in clear disagreement with the data obtained by CW EPR measurements. On the contrary, reasonable N values are obtained when the inversion efficiency is approximated by the modulation depth of a 1:2 HSA–DSA mixture (Table 3.1). Mixtures with 3–5 equivalents of fatty acid give rise to values $N = 2.93$, 3.76, and 4.65, which correlate well with the expected average number of fatty acids per molecule.

The difference in the inversion efficiency of the biradical and the HSA–fatty acid mixtures is far too pronounced to be explained by the slightly narrower EPR spectrum of the biradical (Appendix A.1.9). Further, no angular correlations of the DSA molecules were observed, which would lead to an orientation dependence of λ as indicated by Eq. 2.100 (Appendix A.1.10). This is easily comprehensible since the spin-bearing hydrocarbon ring is still able to rotate rather freely about the methylene chain of the fatty acid, even if the position of the fatty acids is restricted due to narrow protein channels. On the contrary, a slight orientation selection of the spins in the rigid biradical was found (data not shown). While a strong orientation selection would account for the observed deviation in λ [31, 42], the slight orientational dependence observed here does not.

It is well known, that the modulation depth is partly suppressed for dipolar pairs with short distances and the dipole–dipole couplings in the range of or larger than the excitation bandwidth of the pump pulse [41]. But as discussed in detail in the last section, a homogeneous distribution with a main contribution at 3.6 nm is observed, and even the crystallographic data suggests only two distances <2 nm (Table A.1). Additionally, no indications for strong dipolar couplings can be found in ESE detected spectra, as spectral broadenings are absent (Appendix A.1.9). Hence, the reason for the apparent deviation of the inversion efficiency remains unclear.

More realistic values of N are obtained using the 1:2 HSA–DSA mixture as reference. Still, the obtained values are slightly lower than the expected ones with a bigger deviation for large numbers of coupled fatty acids. For an average of 3, 4, and 5 coupled spins, values of 2.93, 3.76, and 4.65 are calculated.

In fact, the modulation depth is slightly overestimated for protein complexes with a high fatty acid content. This is due to the fact that the apparent background dimensionality of 3.8 is based on the assumption of a single electron spin in the center of a sphere with a radius of 2.5 nm (Appendix A.1.5). If more spins are located in the same sphere, the probability of intermolecular distances <5 nm increases and the apparent dimensionality is reduced.

However, this slight overestimation of N is counteracted by a second and larger effect. Bode et al. observed that the contributions from mixtures with a varying number of coupled spins to the overall modulation depth are weighted with a scaling factor which depends on the transverse relaxation time T_2 of the spins in the cluster [39]. Clusters with more coupled spins exhibit a decreased T_2 time, hence their spectral contribution is underrepresented (Appendix A.1.9). On a different note, the height of the central peak of the EPR spectrum is more strongly decreased by the enhanced relaxation times than the flanks. Thus, the inversion efficiency for highly loaded HSA samples is slightly decreased as well (as the pump pulse is located in the spectral center). Since the relaxation time of the isolated spin oligomers could not be accessed, a quantification of this influence according to the method proposed in Ref. [39] was not feasible.

For even higher amounts of spin-labeled fatty acid molecules, the obtained values deviate significantly from the expected values (4.67 vs. 6 and 4.45 vs. 8). In addition to the above mentioned factors, the modulation depth is additionally decreased by contributions from free, uncomplexed spin probes. This effect is most prominent for the 1:8 mixture, which exhibits an even smaller modulation depth than the 1:6 sample. This is in line with CW EPR studies showing that the fraction of unbound DSA is substantially increased for the 1:8 mixture (data not shown). The decrease of the modulation depth due to unbound DSA is also manifested for the spin-diluted systems examined in the last section (Fig. 3.3a). The addition of rDSA molecules leads to an increasing fraction of unbound DSA and to a decrease of the modulation depth.

In conclusion, the modulation depth serves as a means to qualitatively assess the average number of spins in self-assembled systems. A quantitative interpretation is mainly hampered by the existance of spin clusters with a varying number of coupled spins and contributions from unbound spin-labeled material, which decrease the modulation depth. In this study, large deviations are observed for ≥ 5 coupled spins in a total of seven potential binding sites.

3.3.2 *Quantification of Multispin Artifacts*

As already mentioned, the time-domain data in Fig. 3.5a give rise to distance distributions that deviate considerably from the results of the spin-diluted systems for a high degree of loading. While the distance distributions of the spin-diluted systems hardly change with the number of fatty acids (Fig. 3.3b), the distributions with only spin-labeled fatty acids undergo pronounced changes depending on how many DSA molecules are added to the protein (Fig. 3.5b). In addition to a substantial broadening of all distance peaks, their relative ratio is significantly altered. The contribution of the small distance at 2.2 nm increases relative to the dominant contribution at 3.6 nm, while the large distance at ~5 nm vanishes for HSA–fatty acid mixtures with more than three equivalents of DSA.

In the following, flip angle dependent DEER measurements are performed to reveal the influence of multispin contributions to the changes of the distance distributions. The experiment is based on the dependence of N-spin contributions on the inversion efficiency, which can be roughly approximated by

$$V_{N-\mathtt{spin}} \propto \lambda^{N-1}. \tag{3.3}$$

A detailed description of the theoretical background is given in Ref. [32]. Each contribution is affected to a different degree by a variation of the inversion efficiency, which can be controlled by the flip angle of the pump pulse. Hence, a series of measurements at different flip angles allows for the separation of the N-spin interactions. Since three-spin interactions constitute the by far dominant part of the multispin interactions for $\lambda \gg 1$, only the separation of the two- and three-spin contribution is considered in analogy to the method described by Jeschke et al. [32]. As multispin interactions with $N > 3$ are neglected in the analysis, the 1:3 HSA–DSA mixture and the comparison to the spin-diluted samples are discussed in detail.

Intramolecular DEER time-domain signals for six different flip angles of the pump pulse are shown in Fig. 3.6a, b. The flip angles were varied by selective attenuation of the microwave power of the pump pulse channel and the resulting inversion efficiencies were quantified by an inversion recovery sequence. The relation of the modulation depth and the inversion efficiency is given by the relation [32]

$$\Delta(\lambda) = \sum_{i=1}^{N-1} d_i \lambda^i. \tag{3.4}$$

For pure two-spin contributions a linear relationship is expected. Any contribution from multispin interactions gives rise to deviations from this linear dependency due to the admixture of higher order polynomials. Indeed, a slight deviation is observed for 1:3 mixtures of HSA with both 16-DSA and 5-DSA (Fig. 3.6c, d), which is absent if the ratio is decreased to 1:2 (data not shown). This is a clear indication that multispin interactions contribute to the overall DEER signal. The deviation from the linear curve is not as strong as expected for a triradical [32], since the sample contains a mixture of proteins with one, two, three, and higher amounts of spin-labeled fatty acids.

Having visualized the contributions due to multispin interactions, one can now focus on the resulting artifacts in the respective distance distribution. For this purpose, pair and three-spin contributions are extracted from the raw DEER data. In Fig. 3.7, the distance distribution obtained from the pair contribution is

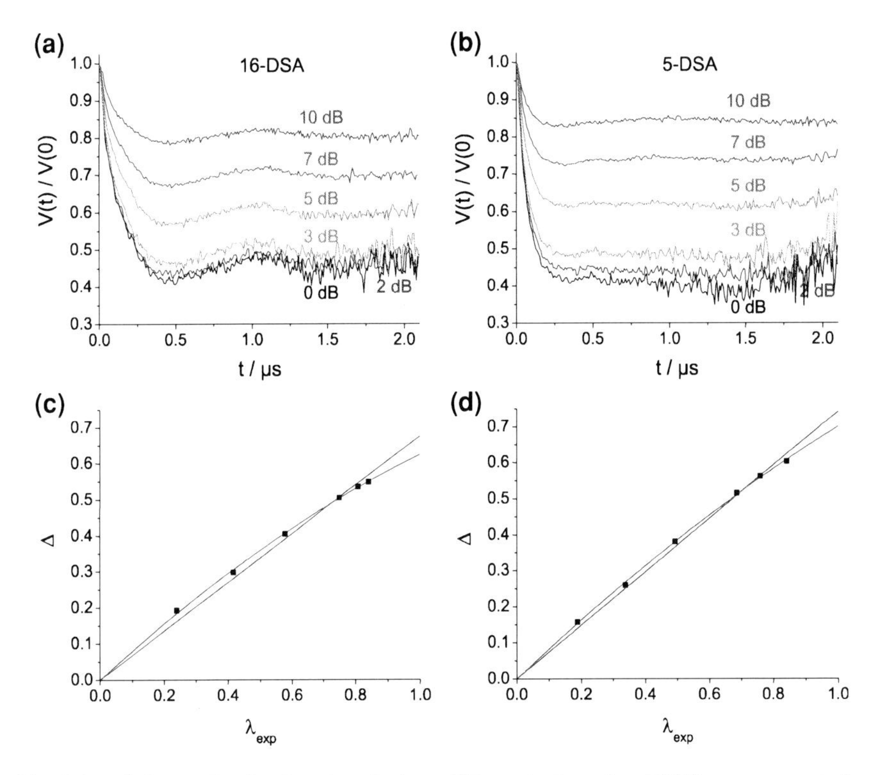

Fig. 3.6 **a, b** Intramolecular time-domain data of flip angle dependent DEER measurements for 16-DSA (*left*) and 5-DSA (*right*). The inversion efficiency of the pump pulse was decreased by attenuation of the microwave power output ranging from 0 to 10 dB. **c, d** Plots of the total modulation depth Δ as function of the inversion efficiency λ_{exp} (determined by inversion recovery). The data points were fitted with a second-order polynomial (*red*) accounting for two- and three-spin contributions to the DEER signal and a straight line, which neglects three-spin contributions. Reprinted with permission from [67]. Copyright 2011 Elsevier Inc

compared to the distribution obtained from the original DEER data, which is distorted by multispin interactions.

The observed changes are most pronounced for 16-DSA, since the distribution contains well-resolved peaks, but are also manifested in the distribution of 5-DSA. The multispin effects, present in the raw DEER data, lead to a slight broadening of the observed distance peaks. More severely, contributions from small distances are overestimated while large distances are suppressed. These observations are in full agreement with the multispin effects observed for model triradicals [32].

With the data at hand, the apparent deviation between the distance distributions of fully loaded HSA samples can solely be related to multispin effects. In a 1:8 mixture of HSA and 16-DSA or in a 1:4 mixture of HSA and 5-DSA, multispin effects lead to an overestimation of small distances and to a broadening of the peaks in full analogy to the flip angle dependent DEER results. Hence, a spin-diluted system is an indispensible prerequisite for retrieving artifact- and

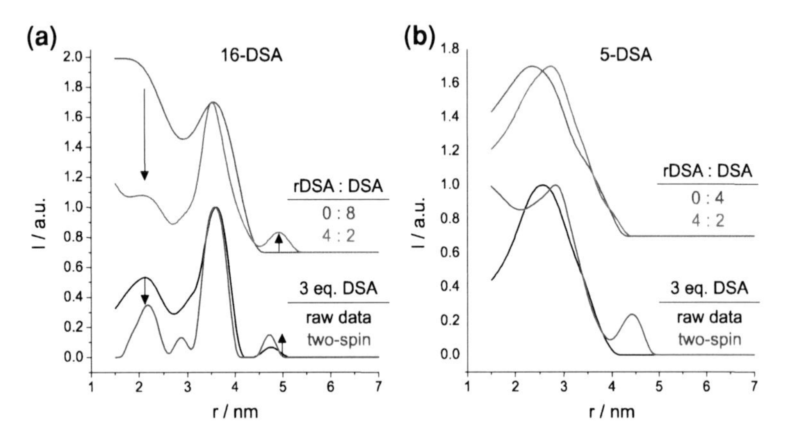

Fig. 3.7 Comparison of the distance distribution obtained from the raw time-domain data of a 1:3 mixture of HSA and DSA (*black*) and from the extracted two-spin contributions (*red*) for 16-DSA (**a**) and 5-DSA (**b**). The observed differences, indicated by *black* arrows for 16-DSA, are compared to the deviations observed for HSA molecules fully loaded with purely paramagnetic DSA (*blue*) and a spin-diluted mixture of rDSA and DSA (*orange*). Reprinted with permission from [67]. Copyright 2011 Elsevier Inc

distortion-free distance distributions when self-assembled systems with a more than two potential paramagnetic centers are studied.

Even for protein samples with a high number of paramagnetic centers, all observed changes are described well by three-spin contributions. It can thus be assumed that contributions from multiple spin interactions with $N > 3$ are negligible, even if as much as seven spins are coupled. Note that, in principle, three-spin contributions can be analyzed in addition to pair contributions. They do not only contain information about the distance, but also about the relative orientation of the three interacting spins [32]. However, this analysis is not carried out in this section due to the plentitude of coupled spins, which generates various possibilities for three spin contributions with different distances and orientations. Instead, orientational information will be accessed in the next section by employing an EPR-active transition metal.

3.4 Orientation Selectivity in DEER: Beyond Distances

Besides fatty acids, HSA binds and transports a large variety of endogenous compounds and drugs. Most drugs are bound in two distinct binding sites located in subdomains IIA and IIIA of the protein [5, 43], which overlap with fatty acid binding sites 2, and 3/4 [11, 12]. Among the endogenous ligands, hemin, bilirubin, and tyroxine are the most important. Hemin and bilirubin bind to subdomain IB (fatty acid site 1) [44–46], whereas thyroxine binds to subdomain IIA (fatty acid site 7) [47].

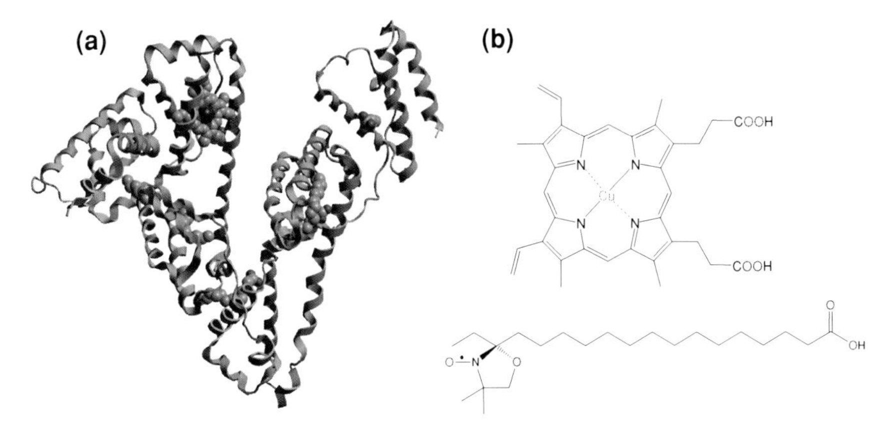

Fig. 3.8 **a** Crystal structure (PDB 1o9x) of HSA co-crystallized with hemin and six myristic acid molecules [44]. **b** Chemical structure of the EPR active substitutes for hemin and myristic acid, Cu(II) protoporphyrin IX and 16-DSA. Reprinted with permission from [68]. Copyright 2011 Biophysical Society

Hemin is a particularly interesting ligand for EPR studies since it contains a paramagnetic Fe^{3+} central ion, which is complexed by a porphyrin derivative. It fits snugly into the hydrophobic cavity of the protein with its two propionate groups interacting with basic side chain residues of the protein [44]. In that sense, its binding mode resembles that of fatty acids. In fact, hemin exhibits a higher affinity to the protein binding site than the fatty acids. Thus, the addition of one equivalent of hemin is sufficient to quantitatively replace the fatty acid bound to site 1. The crystal structure PDB 1o9x of HSA co-crystallized with hemin and myristic acid is shown in Fig. 3.8 [44].

Although Fe^{3+} itself possesses a quadrupolar electron spin with $S = 5/2$, which leads to substantial broadening of the EPR spectrum (Appendix A.1.11), it can easily be substituted by a variety of transition metal ions. Cu^{2+} ($S = 1/2$) is the ion of choice for EPR applications, since its spectra exhibit comparably narrow spectral widths. Due to the relatively high spectral density, Cu^{2+} can even be utilized as a probe in DEER distance measurements (see below, cf. Sect. 4.4) [48, 49].

By the addition of Cu(II) protoporphyrin IX (replacing hemin) and 16-doxyl stearic acid (replacing myristic acid) to HSA, a self-assembled ternary system is obtained that allows a structural characterization of the protein in analogy to Sect. 3.2 (Fig. 3.9). In fact, the spectral separation of copper and nitroxide contributions opens the possibility to solely retrieve distances between the Cu^{2+} ion in the center of the porphyrin and the nitroxide groups of the fatty acids. Thus, the total number of distances probed by the system is significantly reduced from 21 (FA–FA) to 6 (hemin–FA). According to the crystal structure, these six distances range from 2.36 to 4.25 nm (C-16 position of the fatty acids). Hence, they are well in the distance regime that can be accessed by DEER (Fig. 3.9).

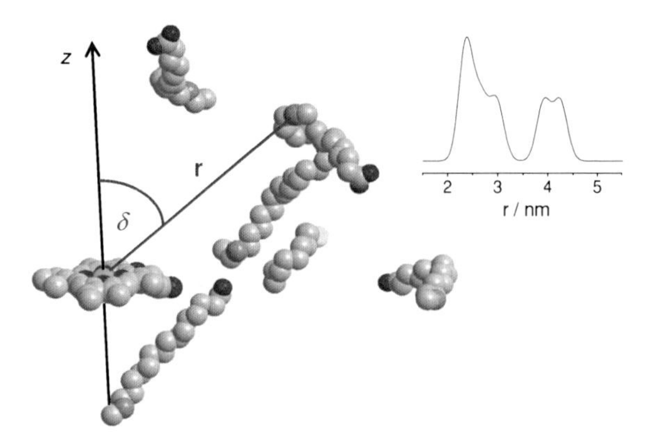

Fig. 3.9 Relative positions of Cu(II) protoporphyrin IX and the stearic acid molecules in the protein. The structural model was obtained by inserting hemin in the crystal structure PDB 1e7i of HSA complexed with seven stearic acid molecules (cf. Sect. 3.6) [8]. The C-16 positions of the stearic acids are highlighted in orange and the last resolved carbon atom of stearic acid in site 7 is colored yellow. The structural relationship between Cu(II) and position C-16 of the stearic acids is sufficiently described by the distance and the angle δ between the connecting vector and the z-axis of the molecular Cu(II) frame. In the inset, the expected distance distribution is shown assuming Gaussian peaks with $\sigma = 0.14$ nm. The methylene chain of the stearic acid in site 7 was linearly extrapolated to its C-16 position. Reprinted with permission from [68]. Copyright 2011 Biophysical Society

Further, the major disadvantage of the selective excitation of only few Cu^{2+} spin packets by a single microwave pulse can be used as an advantage, as the excited spins possess a defined orientation with respect to the external magnetic field. Due to this orientation selection, DEER does not only provide information about the spin–spin distance, but also contains information about the relative orientation of the spin–spin vector with respect to the molecular frames of the paramagnetic centers [50, 51]. Note that the ternary system based on HSA also constitutes a fully self-assembled biological model system for orientation selection in DEER.

The effect of orientation selection on a DEER spectrum can be utilized if the g anisotropy of at least one of the paramagnetic centers is resolved. For nitroxides, sufficient resolution is provided at high magnetic fields of ≥ 95 GHz and spin pair geometries can be accessed [42, 52–55]. However, orientationally selective DEER measurements on nitroxides are also possible at X-band, since a variation of the observer pulse position in the low field part of the spectrum slightly affects the orientation of the excited spins [31, 56].

Transition metals give rise to strong angular effects even at X-band due to the significantly broader g anisotropy. In particular, a variety of Cu^{2+} containing compounds was studied by DEER. Copper–nitroxide distances and orientations were accessed for terpyridine and porphyrin model complexes [48, 57], and even copper–copper distances could be estimated for several biological systems [48, 58, 59].

Several sophisticated methods were developed for the analysis of orientation selection in DEER [42, 55, 56, 60]. A direct conversion of the dipolar data into a distance distribution is not possible, since the orientation dependence of the DEER data is a priori unknown. Hence, a structural model is used to retrieve distance and mutual spin orientations. Based on this model, dipolar couplings are calculated and fitted to the experimental time-domain data or its Fourier transform. All reported methods explicitly calculate the orientation selection of both observer and pump spin. The mutual orientation of the paramagnetic centers in then accounted for by Euler angle transformation of the molecular frame of spin B in the molecular frame of spin A. Finally, the resulting excitation pattern is related to the mutual orientation of spin A to the spin–spin connecting vector.

Hence, this thorough methodology completely describes the mutual orientation of the whole spin system. However, in many cases no defined relationship between both spins exists. This is particularly true for flexible spin labels that can assume a large range of orientations and also applies to the spin labels of the fatty acids in this study, as confirmed by field-swept DEER measurements (Appendix A.1.10). Thus, the orientation selection is almost exclusively governed by the transition metal ion, which exhibits a fixed position with respect to the spin–spin vector and a substantially more pronounced g anisotropy.

In the next paragraphs, a particularly simple method for the analysis of the orientation-selective DEER data is presented, which only considers the orientation-selective excitation of the copper spins.

In the first step, the EPR parameters of the Cu(II) porphyrin system are retrieved by simulation of the copper contribution to the EPR spectrum. With these parameters, the relative fractions of excited spins at each angular orientation with respect to the external magnetic field are calculated with the Easyspin routine 'orisel' for each field position of the observer pulse [34]. Typically, the data are calculated in an angular grid of $\theta \times \varphi = 100 \times 100$, giving rise to 10,000 orientations.

The lengths r of the spin–spin vectors and their angles to the Cu^{2+} molecular z-axis δ were obtained from the crystal structure (Fig. 3.9). The position of the electron spin of the nitroxide was approximated by the C-16 position of the fatty acid. The relative positions of the nitroxides to the Cu(II) porphyrin are fully described by r and δ due to the axial symmetry of the Cu^{2+} molecular frame.

For each orientation, the angle between the external magnetic field and the spin–spin vector θ_{AB}, and the effective $g_{Cu}(\theta)$ value (Eq. 2.46) were determined. These values were then used to calculate the dipolar frequency at each magnetic field position (Eq. 2.101). Dipolar spectra were obtained by weighting the dipolar frequencies (positive and negative) with the calculated fraction of excited spins and by $\sin\theta$ and subsequent summation for all orientations. Time-domain signals were calculated by Eq. 2.100.

This calculation yields DEER data originating from infinitely sharp distance peaks. More realistic data were obtained by assuming a Gaussian broadening of each distance (usually $\sigma = 0.14$ nm). Each peak was subdivided into 40 equally spaced sub-distances with $\Delta r = 0.05$ nm, for which the above-mentioned

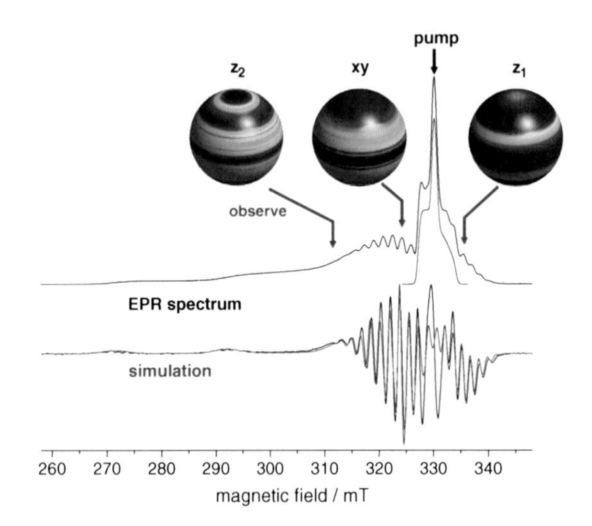

Fig. 3.10 ESE detected EPR absorption spectrum of HSA complexed with 1 eq. Cu(II) protoporphyrin IX and 1 eq. 16-DSA at 10 K (*black, top*). The nitroxide contribution, approximated by a 1:2 mixture of HSA and 16-DSA at 50 K (*blue*), was subtracted. The residual spectrum was pseudomodulated (*black, bottom*) and simulated (*red*). The simulations were used to calculate the Cu(II) orientations (Easyspin function 'orisel') that are excited by a 32 ns pulse at a certain spectral position. In the performed DEER experiments, the frequency of the pump pulse was kept at the maximum of the nitroxide spectrum and in the center of the resonator mode while the position of the observer pulse was varied within the copper spectrum. The orientation selection for these magnetic field positions is displayed in unit sphere plots with warm colors indicating high excitation efficiencies. Reprinted with permission from [68]. Copyright 2011 Biophysical Society

calculation was repeated. To reduce the calculation times, negligible angular and distance contributions were not considered. The cut-off values were 1% (orientation) and 5% (distance) of the maximum value. For six distances, the calculation time was less than 30 min. The source code of the program is given in Appendix A.1.13.

A typical EPR spectrum of HSA complexed with Cu(II) protoporphyrin IX and 16-DSA is displayed in Fig. 3.10 (top). The most prominent feature of the spectrum is due to the nitroxide contribution, which is visualized separately in blue. By subtraction of this contribution, the contribution from the paramagnetic Cu^{2+} was obtained, which could then be subjected to a spectral simulation (red). Uniaxial parameters $g_\perp = 2.053$, $g_\parallel = 2.194$, $A_\perp(\mathrm{Cu}) = 58.8$ MHz, and $A_\parallel(\mathrm{Cu}) = 616$ MHz were retrieved, further superhyperfine couplings to four strongly coupled ^{14}N atoms with magnitudes $A_\perp(\mathrm{N}) = 50.4$ MHz and $A_\parallel(\mathrm{N}) = 37.8$ MHz. These values correspond well to reported values on similar Cu(II) porphyrin systems [61].

For the DEER measurements, the pump pulse was positioned at the maximum of the nitroxide spectrum to minimize orientation selection from these spins and to provide for a large fraction of pumped spins. The position of the observer pulse was varied within the copper spectrum. Orientational unit spheres for three typical

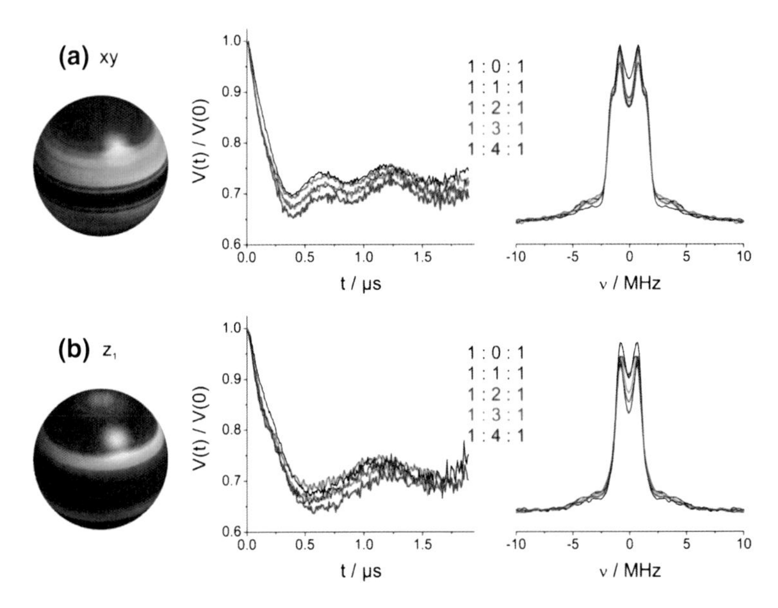

Fig. 3.11 Background-corrected DEER time-domain data and dipolar spectra of HSA complexed with 1 eq. of Cu(II) protoporphyrin IX and 16-DSA, and varying amounts of 16-rDSA, as indicated by the central digit. Observer pulses were applied at field positions exciting Cu(II) orientations either predominantly in the *xy*-plane of the porphyrin ring (**a**, position xy) or outerdiagonal elements towards the molecular *z*-axis (**b**, position z_1). Reprinted with permission from [68]. Copyright 2011 Biophysical Society

positions are depicted in Fig. 3.10. Significant fractions of excited spins along the unique axis of the distorted octahedral frame are only obtained at a low magnetic field. However, the observer pulse position z_2 exhibits a frequency offset of 500 MHz to the pump pulse position at the spectral maximum, which severely decreases the signal-to-noise ratio as the flank of the resonator mode is approached and as the spectral density is decreased. Nonetheless, Bode et al. and Lovett et al., performing DEER measurement on comparable Cu(II) porphyrins, chose this option to achieve the desired orientations along *z* [57, 60]. Luckily, comparable orientations can also be selected at the high field flank of the spectrum (position z_1), which is a result of the strong hyperfine coupling of the strongly coordinated Cu^{2+} ion. As depicted in Appendix A.1.12, both observer pulse positions give rise to similar DEER spectra—with a substantially increased SNR for position z_1.

The obtained DEER time-domain data and frequency spectra for observer pulse positions xy and z_1 are displayed in Fig. 3.11. Note that spin-diluted systems were used in analogy to the fatty acid studies in Sect. 3.2. On average, a protein molecule contains one Cu^{2+} spin and one EPR-active 16-DSA molecule. Occupation of all available fatty acid binding sites is achieved by addition of rDSA. Non-diluted systems give rise to multispin effects as identified in the previous section (Appendix A.1.12).

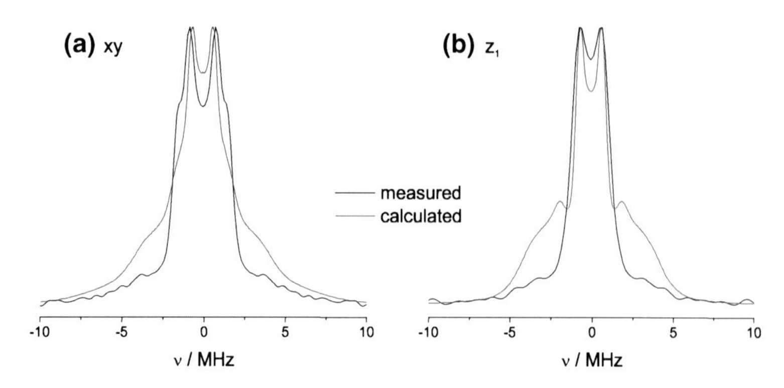

Fig. 3.12 Comparison of the recorded dipolar spectra (HSA complexed with 1:4:1 equivalents of Cu(II):rDSA:DSA) and of the calculated Pake patterns for observer pulse positions xy (**a**) and z_1 (**b**). Full occupation of all fatty acid binding sites 2–7 is assumed. Reprinted with permission from [68]. Copyright 2011 Biophysical Society

Comparing the time traces in Fig. 3.11, a pronounced orientation selection is observed. At the xy position of the observer pulse, the dipolar modulations oscillate with a frequency twice that at the z_1 position. At this position, the dipolar contribution from angles $\theta_{AB} = 0°$ are far more pronounced than in a Pake pattern of a disordered system with no orientational relationship (cf. Fig. 2.23). At the same time, the dipolar singularities at $\theta_{AB} = 90°$ are the only pronounced feature of the dipolar spectrum at an observer position z_1, as contributions from $\theta_{AB} = 0°$ are effectively suppressed.

Remarkably, the dipolar spectra do not undergo considerable changes when the total amount of fatty acid is varied. This observation is in full agreement with the results in Sect. 3.2. It supports the conclusion that the fatty acid sites do either not show pronounced preferences for certain binding sites, or that the binding sites are rather symmetrically distributed within the protein.

One should note that Cu(II) porphyrin does not only serve as EPR probe but also blocks the fatty acid binding site 1. When adding two equivalents of 16-DSA, nitroxide–nitroxide distances can be accessed, which hardly deviate from the data obtained in Sect. 3.2 with seven available binding sites (data not shown). Together with the observations in Sect. 3.2, it can be seen as further indication for a homogeneous, rather symmetric distribution of the entry points to the binding sites in the protein.

In Fig. 3.12, the measured dipolar data are compared to data calculated from the crystal structure, assuming full occupation of all binding sites. The dominating feature of the measured spectra is remarkably well reproduced by the calculated data as concerns both strength and orientation of the dipolar coupling. At the xy position, the enhanced contribution from dipolar angles $\theta_{AB} = 0°$ is clearly visible in the calculated spectra, although it is less pronounced than in the calculated data. In contrast, strong dipolar couplings due to small distances of ~2.5 nm are almost entirely absent in the measured data, though suggested by the crystallographic data.

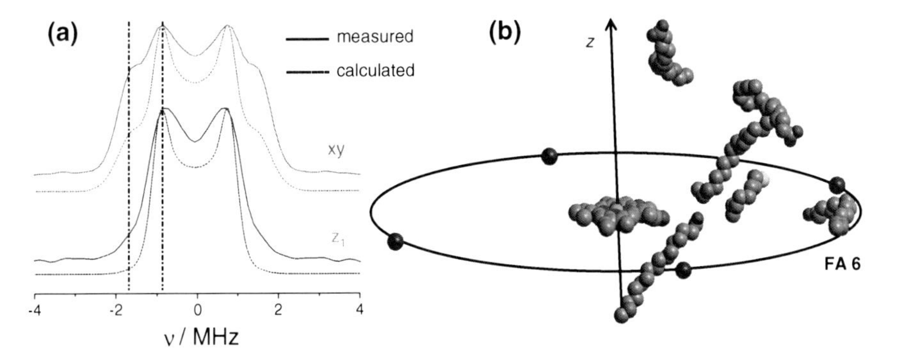

Fig. 3.13 **a** Recorded dipolar spectra (HSA with 1:1:1 Cu(II):rDSA:DSA) and best fit based on a dipolar coupling to a single electron spin placed in the extended xy-plane of the porphyrin ring ($\delta = 90°$) with a distance $r = 3.85$ nm. **b** Structural relation of the dominant DEER contribution to the fatty acid positions of the crystal structure. Potential positions for the dipolar coupled electron spin are indicated by a circle that is located close to the C-16 position of the stearic acid in site 6. Reprinted with permission from [68]. Copyright 2011 Biophysical Society

The dominant contribution in the measured spectra can be unraveled by spectral simulations. Interestingly, this contribution is reproduced by a single dipolar interaction to a coupled nitroxide group (Fig. 3.13a). The qualitative consideration of the last paragraph is confirmed by the quantitative simulation, which yields a distance $r = 3.85$ nm and an angular orientation $\delta = 90°$. Note that the contributions from $\theta_{AB} = 0°$ even exceed the maximum achievable contribution at the given Cu^{2+} orientation selection for the xy observer position.

The obtained parameters correspond to a circle in the porphyrin plane, which is depicted graphically in Fig. 3.13b. Indeed, circular positions almost coincide with the C-16 position of the stearic acid in binding site 6 ($r = 3.82$ nm, $\delta = 88.1°$). This site was already identified as the main contributor to the dominant 16-DSA–16-DSA distance peak by a strict comparison of the EPR data with the crystal structure (Appendix A.1.7).

However, given the fact that the found distance is close to the dominating 16-DSA–16-DSA distance and considering an almost complete absence of other distances, a different explanation is more probable. All observations can be accounted for by a symmetric distribution of the fatty acids tails and the entries to the fatty acid binding sites in the protein, as already proposed in Sect. 3.2. The pronounced orientation selection reflects the fact that the porphyrin ring is roughly aligned to the main plane of the more or less flat protein. Thus, fatty acids are predominantly located at angles close to $\delta = 90°$ as suggested by the crystal structure.

In summary, the orientation selective DEER measurements between the hemin substitute Cu(II) protoporphyrin and 16-DSA confirm the surprising nitroxide distance measurements between two 16-DSA molecules. Again, one dominant dipolar contribution is found, which indicates a symmetric distribution of the binding site entries into the protein.

3.5 Conclusions

DEER and orientationally selective DEER were applied to unravel the functional structure of HSA in solution. It was found that spin-diluted systems are an indispensible prerequisite for artifact-free measurements since multispin effects lead to major distortions of the distance distributions.

From these data one can conclude that HSA in solution is optimized for its function as a transporter of fatty acids. Specifically, the uptake and release of fatty acids is facilitated by a symmetric distribution of the binding sites' entry points. The observed entry point distribution significantly differs from that expected from the crystal structure. While the crystal structure shows an asymmetric distribution of the entry points, a remarkably homogeneous and symmetric distribution is found. Contrary to that, the experimentally derived broad distribution of the ionic head groups as anchoring points shows major similarities with the crystal structure, indicating the reliability of the spectroscopic technique.

In general, the functional solution structures of proteins may differ, even significantly, from the crystal structure. This structure can be accessed by double electron–electron resonance spectroscopy. Using the selectivity of the spin probing approach, only signals from the fatty acids are obtained, which are directly related to the protein's function of interest. This results in a tremendous simplification, but delivers a selective characterization of the functionality of the protein.

3.6 Materials and Methods

Materials. Non-denaturated human serum albumin (HSA, >95%, Calbiochem), 5- and 16-doxylstearic acid (DSA, Aldrich), hemin(chloride) (>98%, Roth), Cu(II) protoporphyrin IX (Frontier Scientific), and 87 wt% glycerol (Fluka) were used as received. The DSA derivatives were partly reduced to EPR-inactive hydroxylamines (rDSA) by addition of phenylhydrazine (97%, Aldrich) as described in detail in Appendix A.1.1.

Sample Preparation. 0.2 M phosphate buffered solutions of pH 6.4 with and without 2 mM HSA and 6.7 mM solutions of DSA and rDSA in 0.1 M KOH were mixed in the appropriate ratios to obtain HSA–fatty acid complexes in a 100 mM phosphate buffer solution of pH 7.4. In Sect. 3.2, the combined concentration of DSA and rDSA was kept constant at 2 mM with varying ratios from 2/0 to 2/6 per protein molecule. This method provides for isolated spin pairs in combination with a varying total occupation of the fatty acid binding sites. In Sect. 3.3, up to eight equivalents of EPR active DSA were added to HSA to study the effect of multispin artifacts in a biological sample. In Sect. 3.4, a ternary mixture of 1 eq. Cu(II) protoporphyrin IX, 1 eq. 16-DSA, and up to 4 eq. of 16-rDSA was complexed with HSA. In addition to already mentioned stock solutions, a 6.7 mM solution of the porphyrin derivative in 0.1 M KOH was prepared for this reason. For DEER measurements, 20 vol% glycerol were added to the final solutions to prevent crystallization upon freezing. The solutions were filled into 3 mm o.d. quartz tubes and shock-frozen in $N_2(l)$ cooled *iso*-pentane (below –100 °C).

Analysis of the Crystal Structure (Sect. 3.2). Distances r_i between the C-16 (or C-5) atoms of the fatty acids in all sites (1–7) were determined from the crystal structure PDB 1e7i (Tables A.1 and A.2 in Appendix A.1.6). The distance peaks were broadened assuming a Gaussian distribution $f(r) = \sum_i \exp\left\{(r - r_i)^2/2\sigma^2\right\}$. Distributions with widths comparable to those of the DEER distributions were obtained with $\sigma = 0.21$ nm (16-DSA) and $\sigma = 0.28$ nm (5-DSA). While two binding modes were suggested for the stearic acid in site 4 [8], the configuration was chosen that is common for all fatty acids and offers electrostatic attachment of the carboxylic acid group. The stearic acid molecules in binding sites 1 and 7 are not fully resolved and were extrapolated to their C-16 position (cf. Appendix A.1.6).

CW EPR Measurements. Continuous wave (CW) EPR spectra were recorded on a Miniscope MS200 (Magnettech, Berlin, Germany) benchtop spectrometer working at X-band ($\sim$9.4 GHz) with a modulation amplitude of 0.04 mT and a microwave power of 50 mW. The temperature was adjusted to 293 K with the temperature control unit TC H02 (Magnettech). No change of the spectra was observed upon addition of glycerol.

DEER Measurements and Analysis (Sect. 3.2). Dipolar time evolution data were obtained at X-band frequencies (9.2–9.4 GHz) with a Bruker Elexsys 580 spectrometer equipped with a Bruker Flexline split-ring resonator ER4118X_MS3 using the four-pulse DEER experiment with the pulse sequence $\pi/2(\nu_{obs}) - \tau_1 - \pi(\nu_{obs}) - (\tau_1 + t) - \pi(\nu_{pump}) - (\tau_2 - t) - \pi(\nu_{obs}) - \tau_2 -$ echo [23, 24]. The dipolar evolution time t was varied, whereas τ_1 and $\tau_2 = 2.5$ μs were kept constant. Proton modulation was averaged by addition of eight time traces of variable τ_1, starting with $\tau_{1,0} = 200$ ns and incrementing by $\Delta\tau_1 = 8$ ns [62]. The resonator was overcoupled to $Q \approx 100$. The pump frequency ν_{pump} was set to the maximum of the EPR spectrum. The observer frequency ν_{obs} was set to $\nu_{pump} + 61.6$ MHz, coinciding with the low field local maximum of the nitroxide spectrum. The observer pulse lengths were 32 ns for both $\pi/2$ and π pulses and the pump pulse length was 12 ns. The temperature was set to 50 K by cooling with a closed cycle cryostat (ARS AF204, customized for pulse EPR, ARS, Macungie, PA, USA). The total measurement time for each sample was around 6 h. The raw time domain DEER data were processed with the program package DeerAnalysis 2008 [41]. Intermolecular contributions were removed by division by an exponential decay with a fractal dimension of $d = 3.8$. The deviation from $d = 3.0$ originates from excluded volume effects due to the size of the protein (see Appendix A.1.5). The resulting time traces were normalized to $t = 0$. Distance distributions were obtained by Tikhonov regularization using regularization parameters of 100 (16-DSA) and 1000 (5-DSA).

Flip Angle Dependent DEER and Data Analysis (Sect. 3.3). The flip angle of the pump pulse β_{pump} was adjusted by the inversion recovery sequence $\beta_{pump} - T - (\pi/2)_{obs} - \tau - \pi_{obs} - \tau -$ echo with $T = 400$ ns and $\tau = 200$ ns on the maximum of the nitroxide spectrum. A maximum inversion efficiency λ_{max} is achieved at a flip angle π_{pump}. The inversion efficiency was defined by $\lambda = 0.5(1 - I_{max}/I_{inv})$ with I_{max} being the echo amplitude without inversion by a pump pulse and I_{inv} the signed amplitude of the inverted echo [32]. The flip angle of the pump pulse was decreased by attenuation (A) of the power output of the external microwave source with a step attenuator DC-18 GHz (Narda Microwave Corporation, New York) to obtain nominal flip angles $\beta = \pi \cdot 10^{-A/20\text{dB}}$ [32]. Attenuator settings of 0, 2, 3, 5, 7, and 10 dB were chosen for each flip angle dependent DEER experiment. The length of the pump pulse was kept at $t_{pump} = 12$ ns to provide for a constant excitation bandwidth. The actual experimental flip angles were calculated by $\beta = \arccos(1 - 2\lambda/\lambda_{max})$ [32]. Dipolar time evolution data for each attenuator setting were obtained with the four-pulse DEER experiment as specified in the previous paragraph. τ_2 was set to 2.2 μs for mixtures of HSA and 16-DSA and to 2.0 μs for samples containing 5-DSA. The measurement time of a single time trace was around 4 h, resulting in a total measurement time of 24 h for each sample. Intermolecular contributions were removed by division by an exponential decay with a fractal dimension of $d = 3.8$. Two-spin and three-spin contributions were extracted from the background-corrected time-domain data with a slightly modified Matlab program kindly provided by Gunnar Jeschke. Details of the data analysis are described in Ref. [32].

Cu(II)–Nitroxide DEER with Orientation Selection (Sect. 3.4). First, an ESE detected spectrum of the combined copper and nitroxide spectrum was recorded with the primary echo sequence $\pi/2 - \tau - \pi - \tau -$ echo with $\tau = 200$ ns. The temperature was set to 10 K. The nitroxide contribution was partly removed by subtraction of a nitroxide spectrum from a 1:2 mixture of HSA and 16-DSA. The residual spectrum was pseudomodulated with a modulation amplitude of 1 mT and simulated with a home-written Matlab program, which utilizes the slow motion routine of the Easyspin software package for EPR (see Sect. 2.7) [34, 63]. Collinear uni-axial **g** and **A** tensors and a natural isotopic composition of Cu (69.2% ^{63}Cu, 30.8% ^{65}Cu) were assumed. Dipolar time evolution data were obtained with the four-pulse DEER experiment (see above). The frequency of the pump pulse was kept at the maximum of the nitroxide spectrum and at the center of the resonator mode while the position of the observer pulse was varied in the copper spectrum (cf. Fig. 3.10).

Simulation of the DEER Spectra. The simulation of the copper spectrum was used to calculate the orientation selection at different observer pulse positions (routine 'orisel' in Easyspin). The orientational weights were implemented in a home-written Matlab program, which simulated DEER time-domain data and frequency spectra on the basis of available crystallographic data. In crystal structure PDB 1o9x, HSA is co-crystallized with hemin and six myristic acid molecules [44], while HSA is co-crystallized with seven stearic acid molecules in

PDB 1e7i [8]. The distances and relative orientations between the central atom of the porphyrin complex (representing Cu^{2+}) and the C-16 positions of the fatty acids (as an approximation for the free nitroxide electron) were obtained by merging the two crystal structures in one common coordinate system. Specifically, hemin was inserted in the crystal structure PDB 1e7i. Its relative position with respect to the fatty acids was calculated utilizing the C-1 atoms of the fatty acids in sites 2–6 as references. The orientational relationship between porphyrin and fatty acid was expressed by the angle δ between the connecting vector (Cu(II)–C-16) and the z-axis of the molecular Cu(II) frame. The C-16 position of the stearic acid in site 7 was obtained by linear extrapolation. The structural relation of Cu(II) to all stearic acids in sites 2–7 gave rise to six distance–angle pairs (r, δ): (2.36 nm, 64.7°), (2.98 nm, 56.2°), (2.38 nm, 85.7°), (4.25 nm, 73.2°), (3.93 nm, 88.5°), and (2.67 nm, 75.8°). The distances were broadened by a Gaussian distribution with $\sigma = 0.14$ nm. Each distance peak was divided into 40 equally spaced contributions ($\Delta r = 0.05$ nm), which were weighted according to the Gaussian distribution.

The molecular coordinate frame of Cu^{2+} served as reference frame, since only copper was assumed to exhibit a pronounced orientation selection. Using Eq. 2.101, 10,000 dipolar frequencies were calculated for each distance contribution of each distance–angle pair, originating from an array of 100×100 orientations of the magnetic field vector in the unit sphere. The effective g-value for each orientation of the magnetic field vector was obtained by Eq. 2.46. The dipolar frequencies were multiplied with the orientational weight for the respective magnetic field orientation, as obtained by 'orisel', and by $\sin \theta$. These frequency contributions were added to generate dipolar spectra and were used to calculate time-domain data according to Eq. 2.100. To shorten the calculation time, only magnetic field vector orientations with weights >1% and distance contributions >5% of the maximum values were considered. One simulation run was completed within 30 min.

MD Simulations. Energy minimization and all MD simulations were performed in the YASARA program package [64]. The crystal structure of HSA co-crystallized with 7 stearic acids (1e7i) was first energy-minimized using the YASARA force field. The energy-minimized protein was put in a $10 \times 10 \times 10$ nm box and filled with water molecules (pH 7.4) to obtain a density of ~1030 g/l. The temperature was set to 298 K. The individual time step was 2 fs and a snapshot of the box was taken every 5 ps. The averaged protein structure obtained of 1200 snapshots (corresponding to relaxation times 0.425–6.425 ns) was analyzed. All distances between the C-16 and C-5 positions of the stearic acids in this structure are summarized in Tables A.3 and A.4 in Appendix A.1.8.

References

1. Peters T (1995) All about albumin: biochemistry genetics and medical applications. Academic, San Diego
2. Carter DC, Ho JX (1994) Adv Protein Chem 45:153–203

3. Spector AA (1975) J Lipid Res 16:165–179
4. Hamilton JA, Cistola DP, Morrisett JD, Sparrow JT, Small DM (1984) Proc Natl Acad Sci USA 81:3718–3722
5. He XM, Carter DC (1992) Nature 358:209–215
6. Curry S, Mandelkow H, Brick P, Franks N (1998) Nat Struct Biol 5:827–835
7. Curry S, Brick P, Franks NP (1999) Biochim Biophys Acta Mol Cell Biol Lipids 1441:131–140
8. Bhattacharya AA, Grüne T, Curry S (2000) J Mol Biol 303:721–732
9. Fasano M, Curry S, Terreno E, Galliano M, Fanali G, Narciso P, Notari S, Ascenzi P (2005) IUBMB Life 57:787–796
10. Hamilton JA, Era S, Bhamidipati SP, Reed RG (1991) Proc Natl Acad Sci USA 88:2051–2054
11. Simard JR, Zunszain PA, Ha CE, Yang JS, Bhagavan NV, Petitpas I, Curry S, Hamilton JA (2005) Proc Natl Acad Sci USA 102:17958–17963
12. Simard JR, Zunszain PA, Hamilton JA, Curry S (2006) J Mol Biol 361:336–351
13. Henzler-Wildman K, Kern D (2007) Nature 450:964–972
14. Lange OF, Lakomek N-A, Farès C, Schröder GF, Walter KFA, Becker S, Meiler J, Grubmüller H, Griesinger C, de Groot BL (2008) Science 320:1471–1475
15. Salmon L, Bouvignies G, Markwick P, Lakomek N, Showalter S, Li D-W, Walter K, Griesinger C, Brüschweiler R, Blackledge M (2009) Angew Chem Int Ed 48:4154–4157
16. Karush F (1950) J Am Chem Soc 72:2705–2713
17. Karush F (1954) J Am Chem Soc 76:5536–5542
18. Laiken N, Nemethy G (1971) Biochem 10:2101–2106
19. Morrisett JD, Pownall HJ, Gotto AM (1975) J Biol Chem 250:2487–2494
20. Rehfeld SJ, Eatough DJ, Plachy WZ (1978) J Lipid Res 19:841–849
21. Livshits VA, Marsh D (2000) Biochim Biophys Acta Biomembr 1466:350–360
22. Milov AD, Salikhov KM, Shirov MD (1981) Fiz Tverd Tela 23:975–982
23. Pannier M, Veit S, Godt A, Jeschke G, Spiess HW (2000) J Magn Reson 142:331–340
24. Jeschke G, Pannier M, Spiess HW (2000) Double electron–electron resonance. In: Berliner LJ, Eaton GR, Eaton SS (eds) Biological magnetic resonance, vol 19: distance measurements in biological systems by EPR. Kluwer Academic, New York
25. Jeschke G (2002) Macromol Rapid Commun 23:227–246
26. Hinderberger D, Schmelz O, Rehahn M, Jeschke G (2004) Angew Chem Int Ed 43:4616–4621
27. Schiemann O, Prisner TF (2007) Q Rev Biophys 40:1–53
28. Dockter C, Volkov A, Bauer C, Polyhach Y, Joly-Lopez Z, Jeschke G, Paulsen H (2009) Proc Natl Acad Sci USA 106:18485–18490
29. Hilger D, Jung H, Padan E, Wegener C, Vogel KP, Steinhoff HJ, Jeschke G (2005) Biophys J 89:1328–1338
30. Schiemann O, Piton N, Plackmeyer J, Bode BE, Prisner TF, Engels JW (2007) Nat Protoc 2:904–923
31. Schiemann O, Cekan P, Margraf D, Prisner TF, Sigurdsson ST (2009) Angew Chem Int Ed 48:3292–3295
32. Jeschke G, Sajid M, Schulte M, Godt A (2009) Phys Chem Chem Phys 11:6580–6591
33. Schneider DJ, Freed JH (1989) Continuous-wave and pulsed ESR methods. In: Berliner LJ, Reuben J (eds) Biological magnetic resonance, vol 8: spin labeling–theory and applications. Plenum Press, New York
34. Stoll S, Schweiger A (2006) J Magn Reson 178:42–55
35. Kawai K, Suzuki T, Oguni M (2006) Biophys J 90:3732–3738
36. Inoue C, Ishikawa M (2000) J Food Sci 65:1187–1193
37. Rariy RV, Klibanov AM (1997) Proc Natl Acad Sci USA 94:13520–13523
38. Milov AD, Ponomarev AB, Tsvetkov YD (1984) Chem Phys Lett 110:67–72
39. Bode BE, Margraf D, Planckmeyer J, Dürner G, Prisner TF, Schiemann O (2007) J Am Chem Soc 129:6736–6745
40. Godt A, Franzen C, Veit S, Enkelmann V, Pannier M, Jeschke G (2000) J Org Chem 65:7575–7582

41. Jeschke G, Chechik V, Ionita P, Godt A, Zimmermann H, Banham J, Timmel CR, Hilger D, Jung H (2006) Appl Magn Reson 30:473–498
42. Polyhach Y, Godt A, Bauer C, Jeschke G (2007) J Magn Reson 185:118–129
43. Curry S (2009) Drug Metab Pharmacokinet 24:342–357
44. Zunszain PA, Ghuman J, Komatsu T, Tsuchida E, Curry S 2003 *BMC Struct. Biol.* 3: 6, doi:10.1186/1472
45. Zunszain PA, Ghuman J, McDonagh AF, Curry S (2008) J Mol Biol 381:394–406
46. Wardell M, Wang ZM, Ho JX, Robert J, Ruker F, Ruble J, Carter DC (2002) Biochem Biophys Res Commun 291:813–819
47. Petitpas I, Petersen CE, Ha CE, Bhattacharya AA, Zunszain PA, Ghuman J, Bhagavan NV, Curry S (2003) Proc Natl Acad Sci USA 100:6440–6445
48. Narr E, Godt A, Jeschke G (2002) Angew Chem Int Ed 41:3907–3910
49. van Amsterdam IMC, Ubbink M, Canters GW, Huber M (2003) Angew Chem Int Ed 42:62–64
50. Larsen RG, Singel DJ (1993) J Chem Phys 98:5134–5146
51. Maryasov AG, Tsvetkov YD, Raap J (1998) Appl Magn Reson 14:101–113
52. Hertel MM, Denysenkov VP, Bennati M, Prisner TF (2005) Magn Reson Chem 43:S248–S255
53. Denysenkov VP, Prisner TF, Stubbe J, Bennati M (2006) Proc Natl Acad Sci USA 103:13386–13390
54. Denysenkov VP, Biglino D, Lubitz W, Prisner TF, Bennati M (2008) Angew Chem Int Ed 47:1224–1227
55. Savitsky A, Dubinskii AA, Flores M, Lubitz W, Möbius K (2007) J Phys Chem B 111:6245–6262
56. Margraf D, Bode BE, Marko A, Schiemann O, Prisner TF (2007) Mol Phys 105:2153–2160
57. Bode BE, Plackmeyer J, Prisner TF, Schiemann O (2008) J Phys Chem A 112:5064–5073
58. Kay CWM, El Mkami H, Cammack R, Evans RW (2007) J Am Chem Soc 129:4868–4869
59. Yang Z, Becker J, Saxena S (2007) J Magn Reson 188:337–343
60. Lovett JE, Bowen AM, Timmel CR, Jones MW, Dilworth JR, Caprotti D, Bell SG, Wong LL, Harmer J (2009) Phys Chem Chem Phys 11:6840–6848
61. Cunningham KL, McNett KM, Pierce RA, Davis KA, Harris HH, Falck DM, McMillin DR (1997) Inorg Chem 36:608–613
62. Jeschke G, Bender A, Paulsen H, Zimmermann H, Godt A (2004) J Magn Reson 169:1–12
63. Hyde JS, Pasenkiewicz-Gierula M, Jesmanowicz A, Antholine WE (1990) Appl Magn Reson 1:483–496
64. Krieger E, Darden T, Nabuurs SB, Finkelstein A, Vriend G (2004) Proteins 57:678–683
65. Junk MJN, Spiess HW, Hinderberger D (2010) Angew Chem 122:8937–8941
66. Junk MJN, Spiess HW, Hinderberger D (2010) Angew Chem Int Ed 49:8755–8759
67. Junk MJN, Spiess HW, Hinderberger D (2011) J Magn Reson 210:210–217
68. Junk MJN, Spiess HW, Hinderberger D (2011) Biophys J 100:2293–2301

Chapter 4
Copper Complexes of Star-Shaped Cholic Acid Oligomers with 1,2,3-Triazole Moieties

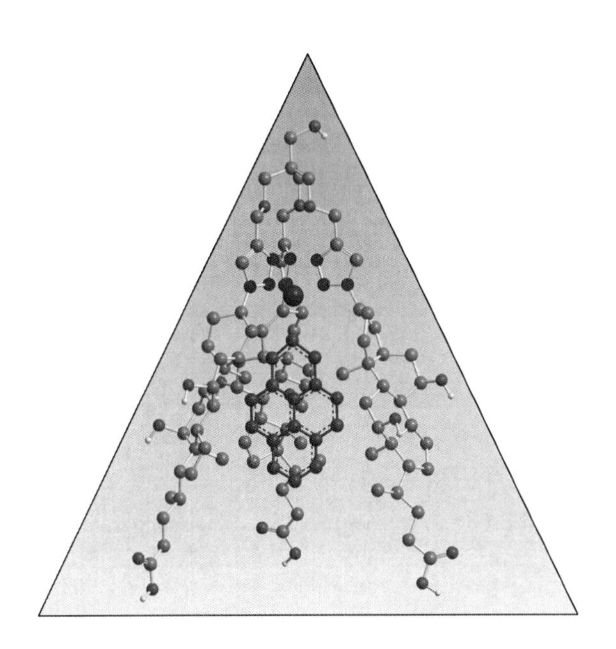

Oligomers based on cholic acid are known to form hydrophobic cavities in hydrophilic solvents due to the facial amphiphilicity of their building blocks. In such pockets, non-polar molecules such as pyrene can be hosted. Oligomers with 1,2,3-triazole moieties are also able to coordinate heavy metal ions. Depending on their position in the cholic acid oligomer, the triazole groups may either cooperatively bind to the metal ion in analogy to a tridentate ligand or act as single entities. This variation of the metal complexation strongly impacts the metal–host interactions. While chelated metals effectively quench the fluorescence of a hosted pyrene molecule, a comparable effect cannot be observed for weakly coordinated

M. J. N. Junk, *Assessing the Functional Structure of Molecular Transporters by EPR Spectroscopy*, Springer Theses, DOI: 10.1007/978-3-642-25135-1_4,

metal ions. Yet, even a weakly ligated metal is able to govern the self-assembly of cholic acid monomers end-capped by two triazole moieties. The spatial distribution of the metal ions in the supramolecular assembly can be observed by DEER spectroscopy.

4.1 Introduction

Bile acids are natural steroid derivatives that exist in humans and most animals. Biosynthesized in the liver, their major function is the emulsification and transport of dietary fats, lipids, and fat-soluble vitamins via micelle formation. Optimized for this purpose, they exhibit a facial amphiphilicity with a hydrophilic concave side due to the presence of hydroxyl groups and a carboxylic group, and a hydrophobic convex side with three methyl groups, triggering their self-assembly and inclusion properties [1].

Due to these characteristics, bile acids are widely used as building blocks in supramolecular chemistry. Materials based on bile acids still exhibit the biocompatibility, amphiphilicity, high stability and self-assembly capacity of the natural precursor. Specifically, various polymers such as pH- and thermo-sensitive polymers [2], star-shaped polymers [3] and crosslinkable monomers [4] have been prepared from bile acids for potential biological and pharmaceutical applications [5, 6]. Bile acids have also been used in the construction of various oligomers, which have been studied as potential drug release agents [7], sensors for metal ions [8], or ion transporters [9].

Among the bile-acid based building blocks, star-shaped derivatives called 'molecular pockets' have drawn much attention. These molecular pockets can form hydrophobic or hydrophilic nano-cavities depending on the polarity of the surrounding solvents. Potential applications include the use as delivery vehicles for biological molecules [10–16], molecular containers and transporters [17–20], and hydrogelators [21, 22].

Beside chenodeoxycholic acid, cholic acid is the most abundant and most important bile acid derivative. In the group of X. X. Zhu, di-, tri-, and tetra-armed star-shaped cholic acid derivatives have been synthesized in the last years [23–25]. These molecules were found to form 'molecular pockets', sheltering hydrophobic molecules such as pyrene in polar solvents or hydrophilic agents such as HPTS (8-hydroxylpyrene-1,3,6-trisulphonic acid trisodium salt) in non-polar solvents [24]. Depending on solvent polarity, a conformational change of the oligomer is induced, which results in an inversion of the hydrophilicity of the pocket (Scheme 4.1) [24, 26]. In polar solvents, the hydrophilic OH-groups turn outwards while they are located inside the pocket in non-polar solvents.

In collaboration with Jiawei Zhang and X. X. Zhu, the structure of triazole-modified cholic acid derivatives was studied via their ability to bind heavy metal ions. For this purpose, cholic acid derivatives containing 1,2,3-triazole moieties were synthesized by a Cu(I) catalyzed 1,3-dipolar cycloaddition, a reaction

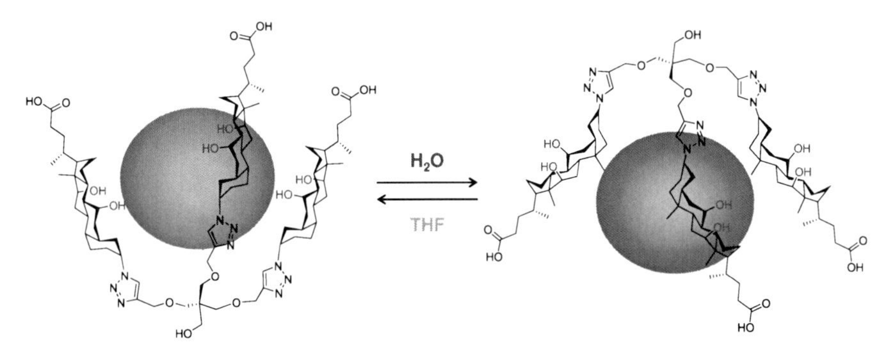

Scheme 4.1 Illustration of the formation and inversion of a 'molecular pocket' of the star-shaped cholic acid derivative T3t. The hydrophilic hydroxyl units are depicted in blue. In THF–water mixtures with up to 15 vol% THF, a hydrophobic cavity is formed, in which non-polar molecules such as pyrene can be sheltered and transported (*right*). Further increase of the THF content results in a conformational change of the oligomer and the formation of hydrophilic pockets (*left*) [24, 25]

commonly referred to as 'click chemistry' [27–29]. In particular, the number of cholic acid arms and the position of the triazole units in the oligomer were varied (Scheme 4.2).

The functional structure of the triazole bearing oligomers was then evaluated by two different methods. The inclusion of the hydrophobic fluorophor pyrene and its interaction with the complexed metal ions was studied by fluorescence spectroscopy using the fact that the polarity of the microenvironment around pyrene is related to the relative ratio of the peaks I_3 to I_1 in the fluorescence spectra [26, 30]. Further, the spatial distribution of the heavy metal ions with respect to pyrene was probed by fluorescence quenching, a photoinduced electron transfer process, which requires molecular contact between fluorophores and quenchers [31].

Complementary structural information from the metal's point of view was accessible via CW and pulse EPR spectroscopy. Using Cu^{2+} as heavy metal ion, the interaction to triazole units was studied in terms of structural arrangement and binding strength. Strong hyperfine couplings to nitrogen atoms manifest themselves as splittings in the EPR spectra while the ligand field strongly affects the electronic structure of copper and thus the **g** tensor [32–35].

This EPR study is not only important as a tool to retrieve structural information about the material of interest. Rather, the cholic acid oligomers, beside their unique properties, also serve as model components for fundamental EPR spectroscopic studies on Cu(II)–1,2,3-triazole interactions, which to my knowledge have never been studied by any means of EPR spectroscopy. Most scientific work was so far focused on its structural isomer 1,2,4-triazole, which was explored as a μ-bridging ligand for the supermolecular assembly of antiferromagnetically coupled Cu–Cu units [36, 37]. Another area of vivid research was related to the binding of copper to imidazole since it is an integral part of histidine and responsible for the binding of heavy metals to some proteins [38–40].

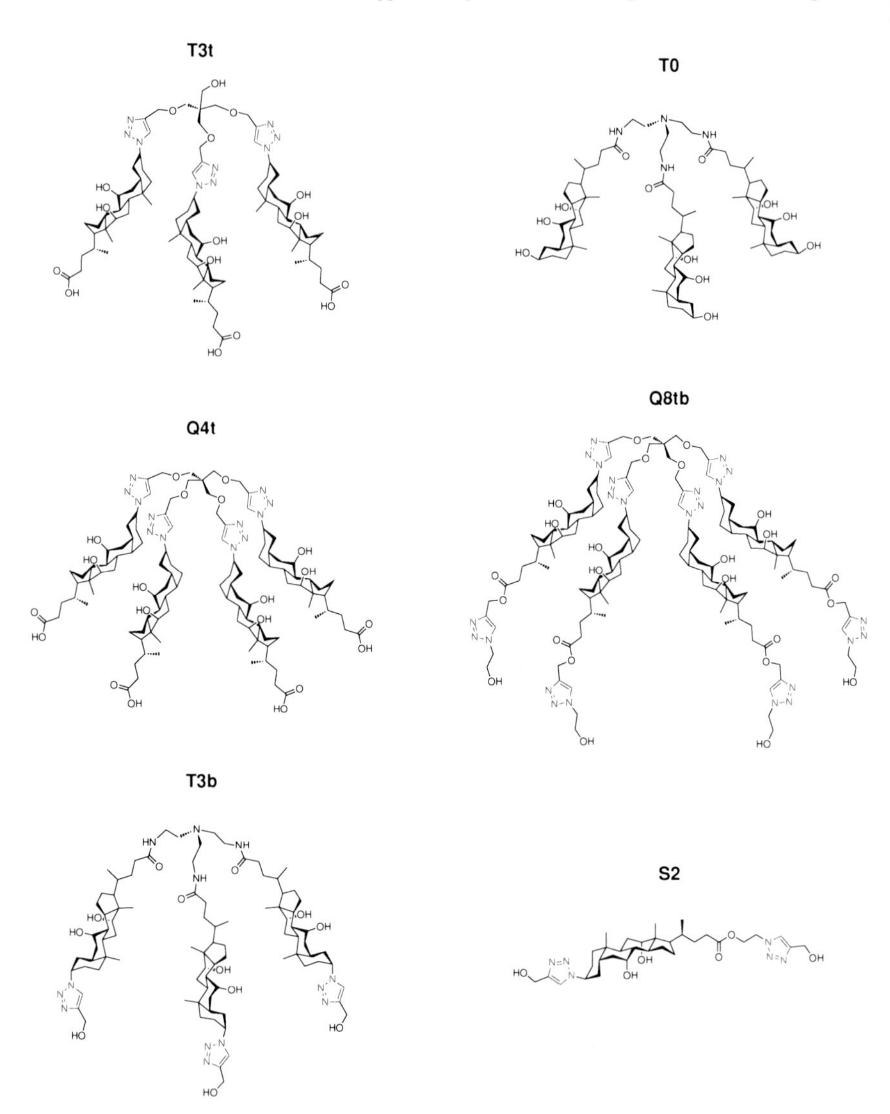

Scheme 4.2 Chemical structures of the studied (star-shaped) cholic acid derivatives. The 1,2,3-triazole moieties are highlighted in orange. The two to four digit abbreviations are structured as follows: The capital letter specifies the number of cholic acid arms in the oligomer (S: monomer, T: trimer, Q: tetramer). The number of triazole groups per molecule is quantified by the digit, followed by the lower case letter, which characterizes their position (t: '*top*', close to the branching unit, and b: '*bottom*', at the tail of the single cholic acid arms)

This chapter is subdivided into three main sections. In the first part, the oligomers T3t and T0 are in the focus. Combined EPR and fluorescence studies reveal that the fluorescence of incorporated pyrene is effectively quenched in T3t. It is shown that this is a consequence of T3t providing a distinct binding site for heavy

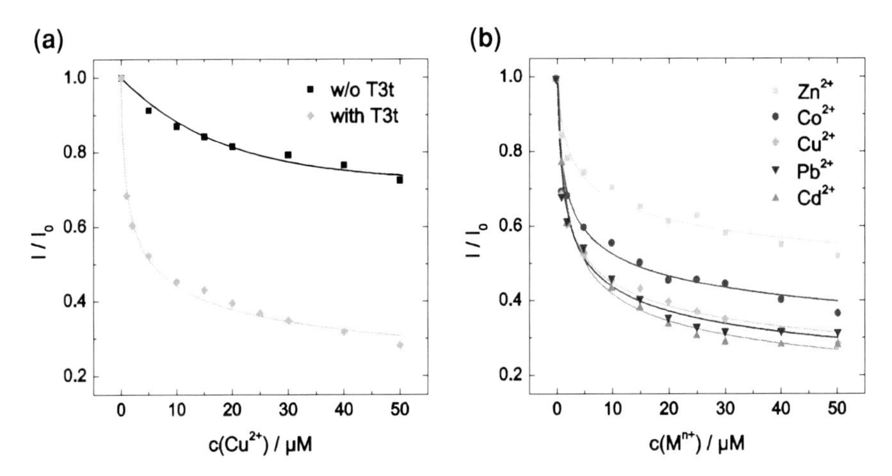

Fig. 4.1 Fluorescence quenching of pyrene (0.1 μM, λ_{ex} = 335 nm) in dependence of the metal ion concentration, measured by the ratio of the fluorescence intensity with (I) and without (I_0) addition of metal ion. **a** Quenching efficiency of Cu^{2+} in the presence (*green*) and absence (*black*) of T3t (25 μM in H_2O). **b** Quenching efficiency of various heavy metals in the presence of T3t (25 μM in H_2O). Reprinted with permission from [25]. Copyright 2010 American Chemical Society

metal ions. The opposite is true for T0, which shelters pyrene from surrounding metal ions. In the second part, the relationship between the molecular structure and metal coordination properties is explored for Q4t, Q8tb and T3b. The binding strength of the copper complexes is then correlated to the observed fluorescence quenching efficiency. Finally, the copper binding and self-assembly of monomeric cholic acid derivatives with two triazole moieties (S2) are examined by CW EPR spectroscopy and DEER.

4.2 Triazole-Substituted Oligomer T3t as Metal Ion Sensor

4.2.1 Fluorescence Quenching Enhancement by T3t

When Cu^{2+} is added to an aqueous solution of pyrene, the original fluorescence intensity I_0 is reduced depending on the amount of added metal ions (Fig. 4.1a, black squares). This reduction of the fluorescence intensity is commonly referred to as quenching and originates from a photoinduced electron transfer process between the fluorophor (pyrene) and the quencher (Cu^{2+}). Upon addition of 50 μM Cu^{2+} to 0.1 μM pyrene, the fluorescence signal is reduced to 70% of its original value.

Remarkably, this quenching effect is exceeded by addition of only 1 μM Cu^{2+} in the presence of T3t (Fig. 4.1a, green diamonds). By fluorescence measurements and utilizing the Benesi-Hildebrand type double reciprocal equations [41], it was found that T3t is able to host one equivalent of pyrene in its hydrophobic pocket [25]. The cholic acid trimer could also act as a receptor for metal ions, since 1,2,3-triazole groups are potential ligands for metal ions, as reported for a triazole-modified calix[4]crown [42]. Hence, the oligomer could provide for a close contact between the metal ion and the pyrene molecule, which is a prerequisite for efficient fluorescence quenching. This metal coordination is studied in detail in the next section by means of EPR spectroscopy.

An enhanced quenching efficiency is observed for a variety of heavy metal ions, as shown in Fig. 4.1b. When 25 μM T3t is added to the solution, the ratio I/I_0 decreases sharply up to a metal ion concentration of 5 μM. Above 25 μM, a plateau is reached with further addition of metal ions only leading to minor changes of the overall quenching efficiency. This suggests a metal–T3t stoichiometry of roughly 1:1, i.e. each oligomer is able to complex one metal ion (cf. EPR results in Sect. 4.2.2). Apparently, the combined complexation of one pyrene molecule and a metal ion by T3t triggers the drastic fluorescence enhancement in presence of the cholic acid oligomer. In fact, a more than 100 times larger quantity of metal ions is required to achieve the same level of fluorescence quenching in the absence of T3t.

In light of this remarkable fluorescence enhancement, T3t can be viewed as a potential sensor for metal ions. Recently, considerable research efforts have been devoted to design and synthesize chemosensors for metal ions [8, 42–50]. In particular, transition metal ions constitute either environmental pollutants or essential trace elements in biological systems. Therefore, reliable detection of these metal ions down to trace amounts is of high importance.

If a decrease of the fluorescence intensity by 20% is used as a criterion of detection, all metal ions studied can already be detected at a concentration of 1 μM with the exception of Zn^{2+} (Fig. 4.2), i.e. the detection limit is at the high ppb level. This limit is better than or comparable with some sensors used in organic solvents [42, 44–46, 50]. In addition to the low detection limit, T3t can be used directly in water and allows for the quantitative analysis of metal ions under environmental conditions.

On a different note, the simultaneous hosting properties of T3t for metals and organic molecules may also be favorable for applications in catalysis. A far less sophisticated Cu–tridentate complex has already been shown to accelerate the hydrolysis of phosphordiesters [51]. Due to its unique binding properties, T3t may also be used as synthetic receptor that bears characteristics of metal-containing enzymes [52, 53].

4.2.2 *T3t: A Tridentate Ligand for Copper*

To obtain information on the complexation of heavy metal ions by triazole moieties of T3t, CW EPR spectra of Cu^{2+} with different admixtures of T3t were recorded (Fig. 4.3). Cu^{2+} was chosen since it does not only exhibit the highest

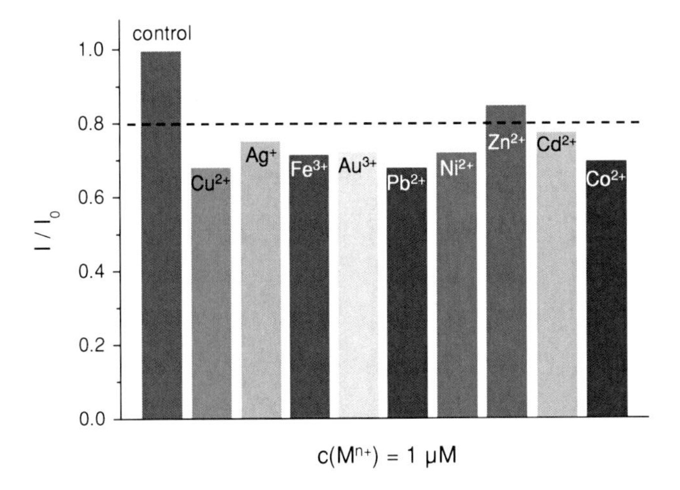

Fig. 4.2 Fluorescence quenching of pyrene (0.1 μM, $\lambda_{ex} = 335$ nm) by the addition of different metal ions (1 μM) in presence of T3t (25 μM in H_2O). The criterion for the detection of a metal ion is set to $I/I_0 = 0.8$ (*broken line*), i.e. when the intensity of the original fluorescence signal is decreased by 20%. Reprinted with permission from [25]. Copyright 2010 American Chemical Society

quenching effects on pyrene but is a commonly used EPR probe due to its effective electron spin of $S = 1/2$ (cf. Sect. 3.4).

When only $CuCl_2$ is present in the solvent (a ternary mixture of ethanol, water, and glycerol), the CW EPR spectrum reflects a uni-axial **g** tensor, which is characteristic of a slightly distorted octahedral coordination of the ion by solvent molecules (Fig. 4.3a). The four well-resolved peaks in the low field region ($B_0 = 258 - 304\,\text{mT}$) contain information about the magnetic interactions along the direction parallel to the unique octahedral axis of the Cu(II) molecular frame. They originate from interactions with the nuclear spins of ^{63}Cu and ^{65}Cu (both $I = 3/2$) with a hyperfine coupling value $A_{\|}$ and are centered around $g_{\|}$. The prominent feature at 322 mT contains information about magnetic interactions in the equatorial (*xy*-) plane and is centered at $g_{\perp}$. Here, four distinct peaks are not resolved since $A_{\perp}$ is too small.

Upon addition of a two-fold excess of T3t to the same system, the spectral features change considerably (Fig. 4.3b). The peaks at low field are less pronounced and shifted to substantially higher B_0-values (corresponding to a decrease of $g_{\|}$) while $A_{\|}$ is increased at the same time. The most prominent changes, however, can be observed in the *xy*-region at 320–340 mT with the formation of a superhyperfine structure with a multitude of sharp spectral features. This hyperfine pattern originates from a strong coupling of the Cu^{2+} ion with ^{14}N atoms and is a first clear indication of an interaction between copper and the triazole units of the oligomer.

When the relative amount of T3t is decreased to an equimolar ratio (Fig. 4.3c), this prominent feature prevails while the low-field region undergoes significant

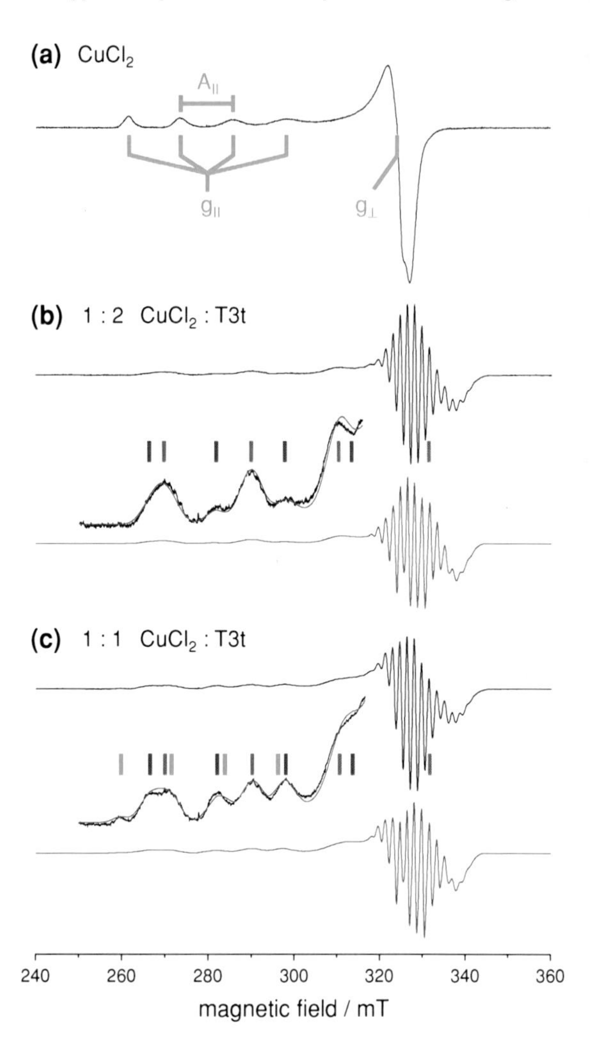

Fig. 4.3 CW EPR spectra of $CuCl_2$–T3t mixtures with ratios of **a** 1:0 (pure $CuCl_2$), **b** 1:2, and **c** 1:1 in a ternary solvent mixture of ethanol, water and glycerol (9:1:2 (v/v)), recorded at 78 K. Spectra are shown in black, the corresponding simulations in red. The low field regions of Fig. 4.3b, c are magnified by a factor of 15. The peak positions of the spectral components parallel to the unique axis of the distorted octahedral Cu(II) frame are indicated by bars: bound (to three triazoles, orange), intermediate (bound to two triazoles, *blue*), free Cu^{2+} species (*green*). Reprinted with permission from [25]. Copyright 2010 American Chemical Society

changes. The peak at 269.8 mT is considerably broadened and deviates strongly from a Gaussian shape and two new peaks at 281.9 and 297.9 mT (mere shoulders in Fig. 4.3b) arise. In fact, these peaks originate from a second copper species with different $g_{\|}$- and $A_{\|}$-values. Additionally, a small but distinct bump is visible at 259.6 mT, which coincides in position and shape with the first low-field peak of the reference spectrum of pure $CuCl_2$ (Fig. 4.3a) and can be assigned to Cu^{2+} ions that are not coordinated to triazoles. In conclusion, three different copper species are observed in $CuCl_2$–T3t mixtures with their relative contributions to the EPR spectrum depending on the ratio of the metal ion and the cholic acid oligomer.

To quantify the EPR parameters of each species and their relative contributions to the EPR spectrum, spectral simulations were performed. The spectrum in Fig. 4.3a solely originates from free Cu^{2+}. Its spectral simulation is shown in

Table 4.1 EPR parameters of the observed copper species in T3t

Cu^{2+} species	$g_\perp$	$g_\parallel$	$A_\perp$(Cu)/MHz	$A_\parallel$(Cu)/MHz	$A_\perp$(N)/MHz	$A_\parallel$(N)/MHz	Triazoles coupled
Free	2.084	2.419	28.0	403.6	-	-	0
Intermediate	2.063	2.323	44.0	514.0	46.2	45.0	2
Bound	2.054	2.241	59.8	623.0	48.3	47.6	3

Table 4.2 Spectral contribution the observed copper species for different Cu–T3t ratios

Ratio	Relative spectral contribution		
$CuCl_2$: T3t	Free	Intermediate	Bound
1:2	-	0.18 ± 0.05	0.82 ± 0.05
1:1	0.10 ± 0.02	0.32 ± 0.05	0.58 ± 0.05

Appendix A.2.3. The spectral separation of both triazole-complexed species was achieved by subtracting spectra of different $CuCl_2$–T3t mixtures in the appropriate ratios (Appendix A.2.1). The recorded spectra were then reproduced by a linear combination S_{tot} of the (normalized) simulated spectra of the three different species S_i (Fig. 4.3 b, c, red curves),

$$S_{\mathrm{tot}} = \sum_{i=0}^{3} x_i S_i. \tag{4.1}$$

All EPR parameters of the copper species are listed in Table 4.1. Their relative contributions to the EPR spectra of different $CuCl_2$–T3t mixtures are listed in Table 4.2.

A first important result can be retrieved from the ^{14}N hyperfine lines. The mere appearance of these lines reveals coupling to triazole units. Further, from the number and intensities of these lines, the number of coupled triazole units can be deduced. The major component in Fig. 4.3b (orange bars) is coupled to three triazoles and is referred to as 'bound' species from now on. The minor component (blue bars) is coupled to two triazole units and exhibits slightly lower coupling constants to nitrogen and is referred to as 'intermediate' species.

These conclusions regarding the number of coordinated triazole units are corroborated by the observed g-values. As stated in Sect. 2.3.1, the deviation of the principal **g** tensor elements from g_e mainly depends on the (inherently negative) spin orbit coupling constant λ and on the energetic difference of the SOMO to the highest fully occupied or lowest unoccupied molecular orbital [54]. For Cu^{2+} with a d^9 configuration and a distorted octahedral coordination sphere, the unpaired electron is located in the energetically highest d orbital, $d_{x^2-y^2}$. Neglecting the influence of covalent mixing, the g-values can be approximated by [55]

$$g_\parallel = g_e - \frac{8\lambda}{E(d_{x^2-y^2}) - E(d_{xy})}, \tag{4.2}$$

$$g_{\perp} = g_e - \frac{2\lambda}{E(d_{x^2-y^2}) - E(d_{xz}, d_{yz})}. \tag{4.3}$$

Triazoles as strong ligands induce a large ligand field (compared to water and ethanol). Hence, the energy difference between the *d* orbitals is increased, which results in a decrease of the g-values. When the strength of the ligand coordination remains constant, the reduction is determined by the number of coupled ligands. This trend is observed for the different spectral species (Table 4.1). While the highest g-values are observed for the uncoordinated copper species, they are significantly reduced for the intermediate and even more strongly decreased for the bound species. This decrease of the g-values is accompanied by a concomitant increase of the hyperfine coupling to the copper nucleus.

The ^{14}N hyperfine coupling parameters of the intermediate and bound species are in the range of $a_{iso} \sim 45.8–48.1$ MHz, comparable to those of Cu(II)-porphyrin systems. For a variety of magnetically diluted porphyrins, a value of $a_{iso} = 48.0$ MHz was reported [35]. In this thesis, the ^{14}N hyperfine coupling of a Cu(II)-protoporphyrin complex in human serum albumin was determined as 46.1 MHz (cf. Sect. 3.4). This similarity of the coupling parameters is remarkable, bearing in mind that porphyrin is a tetradentate ligand, which is optimized for metal complexation.

Due to the comparability of the Cu–nitrogen binding strength, the g-values of Cu(II) porphyrins set a benchmark for a complexation by four ligands. Independent of solvent conditions, their $g_{\parallel}$-value was determined reproducibly as 2.18–2.20 [32, 35, 56], in agreement with $g_{\parallel} = 2.194$, which was found in this thesis (cf. Sect. 3.4). This value is slightly lower than $g_{\parallel}$ of the bound species and confirms that not four but three triazoles are coordinated to this copper ion. Further, the g- and A-values of the intermediate species are closer to those of the bound species than to the corresponding values of the solvent-ligated, free Cu^{2+} ion. This is an additional indication, that two nitrogen atoms, i.e. triazole units, are coupled. Remarkably, no copper species is observed which coordinates to only one triazole group. This will be discussed in detail in Sect. 4.3.

Regarding the geometrical structure of T3t and the strong hyperfine coupling of three nitrogen molecules, it can safely be concluded that the ‘bound’ copper species is located in the T3t cavity and chelated by all three triazole groups of the molecule. The ‘intermediate’ species, which exhibits slightly lower coupling constants to nitrogen and which is complexed by only two triazole units, could be either located in the cavity (only coordinated by two triazoles), or it could interconnect two different cholic acid trimers as a chemical crosslinker. This question cannot be solved with the EPR data presented up to this point. Yet, it will become evident in the next section that a strong chelating effect of the triazole ligands is needed to induce a hyperfine coupling, which is as strong as the one observed here. This can only be caused by the intramolecular contribution of two triazole units that are structurally optimized for the complexation. Hence, the ‘intermediate’ species is most likely located inside a cavity which is formed by the self-assembly of two cholic acid arms of the oligomer while the third arm is not involved in the formation of a molecular pocket.

In an equimolar mixture of $CuCl_2$ and T3t, already 90% of all copper ions are coordinated to the cholic acid derivative while only 10% remain uncoordinated (Table 4.2). This observation is supported by the fluorescence quenching studies in Sect. 4.2.1. At this ratio, almost 2/3 of these triazole-ligated ions are placed inside the hydrophobic cavity, which is formed by all three cholic acid arms of the trimer ('bound species'). A minority of 33% is placed in a molecular pocket formed by only two cholic acid arms ('intermediate' species). Upon addition of a two-fold excess of T3t, already 82% of all Cu^{2+} ions are coordinated by all three triazole groups of T3t, while only a remainder of 18% experiences the coupling to only two triazole units. At this Cu–T3t ratio, no 'free', uncoordinated species is observed.

4.2.3 Deriving a Molecular Picture from the EPR Results

As clearly shown by the CW EPR studies, the majority of the metal ions is strongly bound in a cavity that is formed by all cholic acid chains of the three-armed molecule even at an equimolar ratio of copper and oligomer. In all fluorescence quenching studies, only a small fraction of metal was added to a large excess of the cholic acid derivative. Following the observed trend (Table 4.2), it can be safely concluded that almost all metal ions are in this 'bound' state.

From the fluorescence data it is apparent that pyrene molecules enter the hydrophobic cavity of T3t as well. Previous lifetime studies of different trimers at 25 μM indicated that a large fraction of the hydrophobic guests (83–87%) is located inside the cavities formed by the host molecules, while a only small amount (13–17%) remains outside [23].

Hence, T3t is able to complex one pyrene molecule as well as one metal ion and hosts both species as guests inside its hydrophobic cavity as illustrated in Fig. 4.4. Due to this structural arrangement, the metal ion is located in close proximity to the pyrene molecule. This fixed molecular contact is the key to a very efficient fluorescence quenching process that permits the detection of metal ions down to the ppb range in presence of T3t.

4.2.4 Comparative Fluorescence Quenching Studies of T3t and T0

While T3t has been shown to chelate heavy metal ions due to the presence of triazole moieties, T0 should not be able to form complexes with metal ions. While largely similar in structure to T3b or T3t, T0 does not feature potential chelating triazole groups. Thus, T0 should exhibit a substantial decrease of the fluorescence quenching efficiency in comparison to T3t. In fact, the quenching efficiency is dramatically reduced by several orders of magnitude and even far below the case

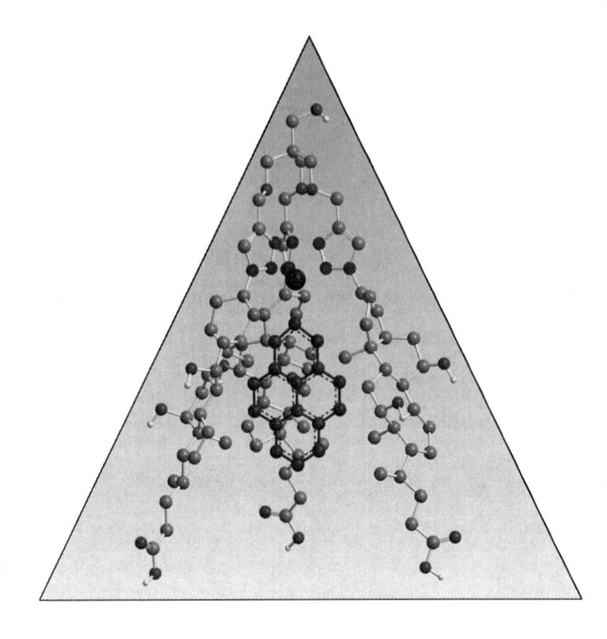

Fig. 4.4 Three dimensional representation of the complex formed by T3t, pyrene, and a heavy metal ion. The metal ion (*green*) is chelated by three triazole units and is located in close proximity to pyrene (*orange*), which is assumed in the center of the hydrophobic cavity

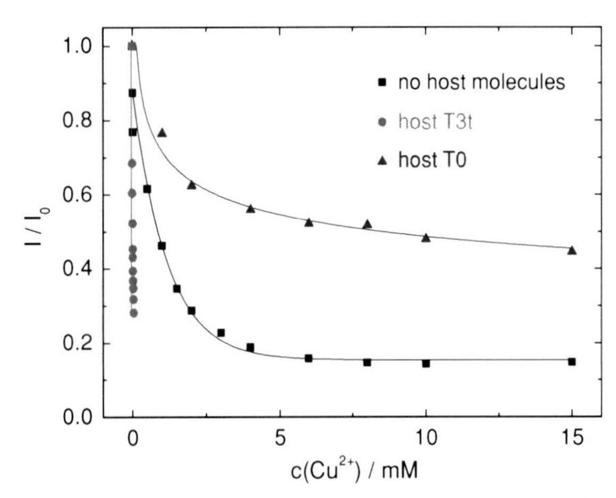

Fig. 4.5 Fluorescence quenching of pyrene (0.1 μM, λ_{ex} = 335 nm) by Cu^{2+} in the absence of a host molecule (*black squares*) and in the presence of 25 μM T3t (*green circles*) and T0 (*blue triangles*) in H_2O. Reprinted with permission from [25]. Copyright 2010 American Chemical Society

when no host molecules were added to the solution (Fig. 4.5). In presence of T0, 15 mM Cu^{2+} ions are needed to decrease the fluorescence intensity to 60% of its original value, while a concentration of only 1 mM is required in the absence of any host molecule. In comparison, already 2 μM Cu^{2+} ions induce a comparable signal reduction in presence of T3t.

Indeed, this diminished quenching suggests that, in contrast to T3t, T0 effectively separates the metal ion quenchers from the pyrene fluorophors. Since both T0 and T3t display comparable hosting properties for pyrene [23, 25], one can conclude that metal ions are excluded from the molecular pocket due to the lack of ligating units. Thus, they are efficiently separated from pyrene, which is located inside the cavity. A molecular contact between both molecules is restricted and, consequently, the fluorescence quenching is decreased.

4.2.5 Summary

In summary, both star shaped cholic acid trimers T3t and T0 form hydrophobic cavities in hydrophilic solvents, which can host non-polar molecules such as pyrene. Containing 1,2,3-triazole moieties, T3t also acts as a strong tridentate ligand for heavy metal ions, which are located in close proximity to the sheltered pyrene molecules. Thus, an efficient fluorescence quenching is facilitated even for low metal concentrations. Due to its sensitivity and water solubility, T3t can be used as a sensor for heavy metal ions in aqueous environments. In contrast, due to the lack of potential ligands, T0 shields pyrene from the influence of metal ions and causes a diminished fluorescence quenching.

4.3 Influence of the Molecular Structure on the Metal Ion Complexation

4.3.1 Fluorescence Quenching of T3b

In T3b, the triazole moieties are located at the tail of the cholic acid arms (Scheme 4.2). In analogy to T3t and T0, it forms a hydrophobic cavity in aqueous solution, in which pyrene can be incorporated. Although its hosting characteristics are slightly less pronounced than those of T3t (Appendix A.2.2), the modified position of the triazole groups hardly influences the uptake of pyrene. Also, a conformational change of the oligomer is induced in response to the solvent polarity, which results in an inversion of the cavity. In a THF–water mixture, this critical solvent composition was determined as 1:6 (v/v) for all three cholic acid trimers, T0, T3t, and T3b [57]. Hence, all structural and functional features that originate from the facial amphiphilicity of the cholic acid building blocks are not influenced by the incorporation and position of the triazole units.

With these triazole groups at the tails of the cholic acid chains, T3b bears the same heterocyclic moieties that were responsible for the strong coordination of metal ions to T3t and lead to an efficient fluorescence quenching. However, the Cu^{2+} induced fluorescence quenching of the T3b–pyrene complex is much less efficient than that of pyrene hosted by T3t (Fig. 4.6). Indeed, the quenching efficiency of the T3b–pyrene complex rather resembles that of free pyrene without addition of any host molecule. Since the inclusion properties of both molecules are comparable, the different quenching efficiencies originate from an altered coordination of the metal ion by the triazoles. The characteristics of this modified complexation are studied in detail in the next section.

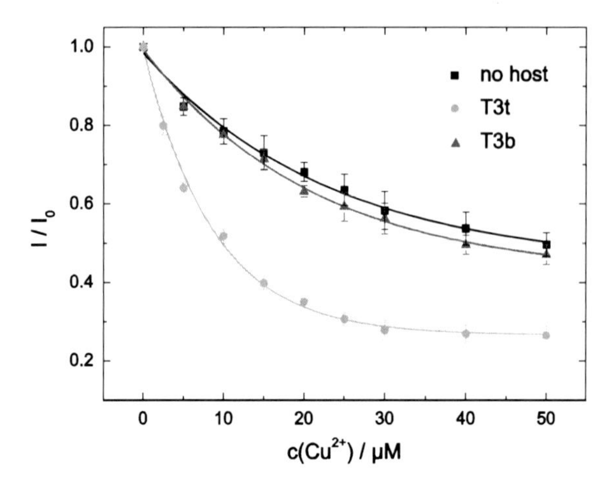

Fig. 4.6 Fluorescence quenching of pyrene (0.1 µM, λ_{ex} = 335 nm) by Cu^{2+} in the absence of a host molecule (*black squares*) and in the presence of 25 µM T3t (*green circles*) and T3b (*red triangles*) in H_2O admixed with 5 vol% methanol. Note that the quenching efficiencies in presence of T3t and in absence of any host are altered in comparison to Fig. 4.1 due to the slight change of the solvent composition. Reprinted with permission from [68]. Copyright 2010 American Chemical Society

4.3.2 Copper Coordination in T3b

CW EPR spectra of Cu–T3b mixtures are displayed in Fig. 4.7. Obviously, no characteristic ^{14}N superhyperfine couplings are observed in the high field region of the spectra. This is in strong contrast to T3t, where these couplings constitute the most prominent feature of the spectrum. Furthermore, it is a first indication that the coordination of Cu^{2+} by triazoles of T3b is either too weak or too undefined to be resolved in a CW EPR spectrum.

A more detailed analysis reveals that three different Cu^{2+} species contribute to the spectra, which is in analogy to T3t. But while free, uncoordinated copper is present in all mixtures of Cu^{2+} and the cholic acid derivatives irrespective of the structure of the oligomer, the other two spectral contributions deviate significantly from those observed in T3t. However, they coincide with the species observed in mixtures of the triazole-modified cholic acid monomer S2. The deconvoluted contributions of each species and the corresponding spectral simulations are displayed in Fig. 4.7, and the extracted EPR parameters are listed in Table 4.3.

These species, denoted 'loosely bound1' and 'loosely bound2', originate from Cu^{2+}, which is weakly coordinated by triazole units. The weak coordination manifests itself in the slight decrease of $g_{\|}$ and in the slight increase of $A_{\|}$ compared to free Cu^{2+}. The values of the 'loosely bound2' species coincide with those of the 'intermediate' species, while the values of the 'loosely bound1' species are in between those of 'intermediate' and 'free' Cu^{2+}. For both species, the ligand

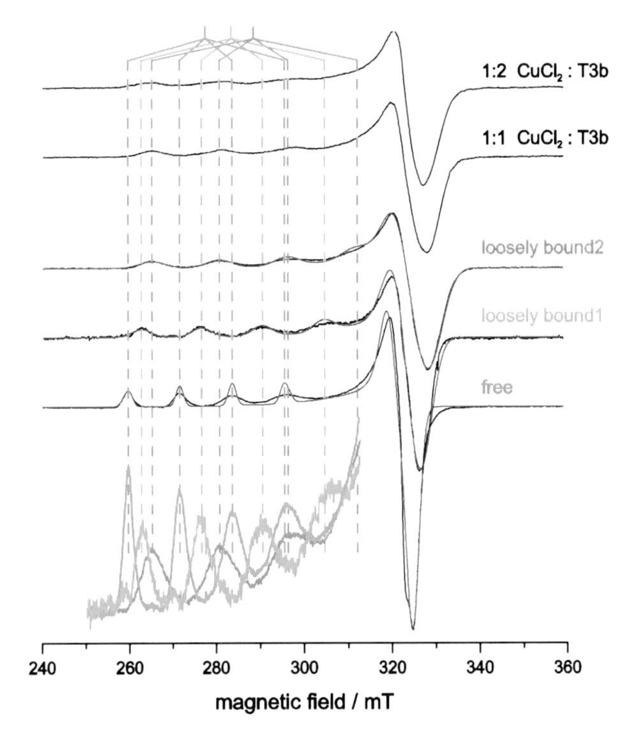

Fig. 4.7 CW EPR spectra of 1:1 and 1:2 mixtures of $CuCl_2$ and T3b (*top*). The spectra consist of three spectral contributions, 'free', 'loosely bound1', and 'loosely bound2'. The isolated spectra of these species are shown below in black, superimposed by their simulations (*red*). The spectra of both 'loosely bound' species were obtained by linear combinations of $CuCl_2$–S2 spectra (which exhibit the same Cu^{2+} species, cf. Sect. 4.4): R(ib1) = 1.044 S2(2:1) – 0.628 S2(1:1)–0.22 S2(1:4) and R(ib2) = S2(1:4) – 0.2 S2(2:1) + 0.082 S2(1:0). All spectra and simulations were normalized to their double integral. Vertical lines mark the $g_{\|}$ transitions of the free (*green*), 'loosely bound1' (*turquoise*), and 'loosely bound2' (*ochre*) species. The low field regions of all spectral contributions are magnified by a factor of 10. Reprinted with permission from [68]. Copyright 2010 American Chemical Society

Table 4.3 EPR parameters of the observed copper species in T3b (and S2)[a]

Cu^{2+} species	$g_{\perp}$	$g_{\|}$	$A_{\|}$/MHz	Triazoles 'coupled'
Free	2.084	2.419	403.6	-
Loosely bound1	2.077	2.370	453.6	1
Loosely bound2	2.072	2.330	498.4	2

[a] $A_{\perp}$ could not be retrieved from the simulations due to lack of spectral resolution

field experienced by the metal ion is significantly decreased when compared to Cu^{2+} bound to three triazole moieties.

The magnitudes of $g_{\|}$ and $A_{\|}$ are first indications for the number of coupled triazoles. Since the parameters for species 'loosely bound2' coincide with those of the species 'intermediate', a coordination to two triazole units is likely. Species 'loosely bound1' experiences a weaker ligand field, which most likely originates from a coupling to one triazole. Hence, the suffixes '1' and '2' relate to the number of coupled triazole units. On the other hand, the term 'loosely bound' is chosen, since it characterizes the rather undefined coordination of the metal by triazole ligands and its decreased binding strength. This conclusion is supported by observations from pulse and CW EPR spectroscopy and fluorescence spectroscopy.

3-pulse ESEEM and HYSCORE are one- and two-dimensional methods to resolve weak electron–nuclear hyperfine couplings (cf. Sect. 2.10) [58, 59]. Representative spectra for $CuCl_2$–T3b and $CuCl_2$–T3t mixtures are shown in Fig. 4.8. The spectra exhibit marked differences, which are explained in detail in the next paragraphs.

The spectrum of $CuCl_2$–T3b contains a multitude of distinct resonances in the (−, +) quadrant of the spectrum, while these resonances are almost completely absent for the $CuCl_2$–T3t complex. Resonances in this quadrant originate from strongly coupled nuclei with hyperfine couplings greater than the nuclear Larmor frequency. Spectral simulations suggest that these resonances are due to a ^{14}N atom with a maximum hyperfine coupling of ~6 MHz (cf. Sect. 4.4.1). This ^{14}N atom is assigned as the part of the triazole group, which acts as direct ligand of the Cu^{2+} ion. For T3t, this directly coupled nitrogen nuclear spin exhibited a hyperfine coupling constant of ~48 MHz, which is strong enough to result in superhyperfine features in the CW EPR spectrum. Such large couplings cannot be observed in HYSCORE spectra. Hence, in the absence of a chelating effect, the Cu–N coupling strength is decreased by a factor of ~8.

Both spectra contain features that originate from at least one weakly coupled ^{14}N nucleus near the exact cancellation limit $|A/2| \cong |\omega_I|$. These remote triazole nuclei give rise to the two prominent matrix peaks in the (+, +) quadrant at 0.92 and 1.83 MHz and to a broader feature at ~4 MHz. These resonances also constitute the main features of 3-pulse ESEEM spectra (Appendix A.2.4). The observed frequencies are similar to those found for the remote ^{14}N atom of an imidazole group. Here, marked resonances at ~0.8, 1.6, and 4.2 MHz were observed and assigned to the nuclear quadrupole transitions ν_-, ν_+ and ν_{dq} in the condition of exact cancellation [60, 61].

In addition, the $CuCl_2$–T3t complex exhibits further matrix peaks up to a frequency of ~6 MHz. As suggested by spectral simulation, these additional peaks could originate from strong nuclear quadrupolar couplings. However, regarding the molecular structure that was unambiguously determined by the CW spectra, these resonances are more likely interpreted as multiquantum transitions. The probability of such transitions increases when a large number of in this case weakly coupled nuclear spins is complexed to an electron spin. In case of T3t, the Cu^{2+} ion is ligated by three triazole groups, which contain six remote nitrogen atoms. In the framework of multiquantum transitions, the absence of these resonances in case of T3b can be easily explained, since only one or two triazoles are coupled to Cu^{2+} in a more undefined manner. Further indications are given by the

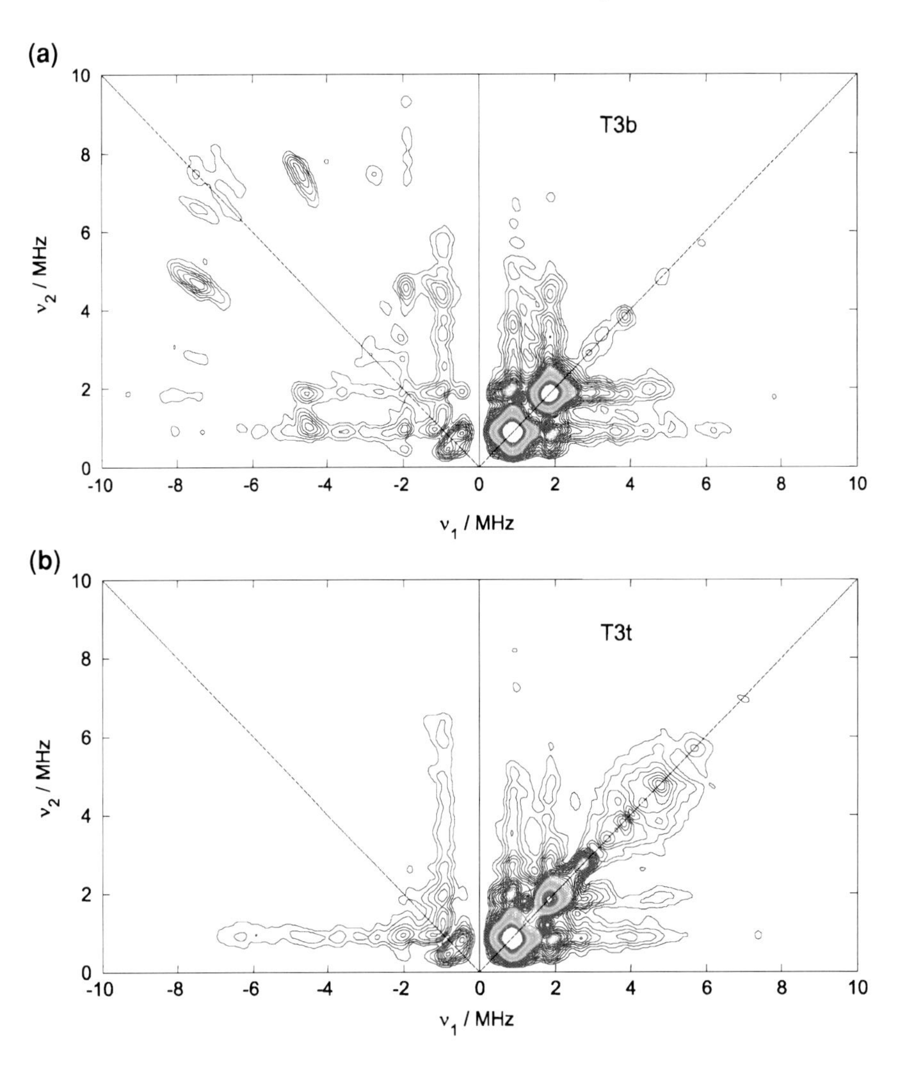

Fig. 4.8 HYSCORE spectra of 1:2 mixtures of **a** $CuCl_2$–T3b, and **b** $CuCl_2$–T3t. The spectra were recorded at the EPR spectral maximum (corresponding to $g_\perp$ **b**) and at a 25 mT lower magnetic field position (**a**). Note that T3t did not exhibit resonances from strongly coupled ^{14}N atoms at any spectral position. The spectra were recorded at 10 K, τ was set to 132 ns. Reprinted with permission from [68]. Copyright 2010 American Chemical Society

equal spacing of the resonances and by the fact that frequencies close to multiples of the ^{14}N Larmor frequency are observed. Finally, it should be noted that the triazole moieties are the only residues of the cholic acid oligomers that act as ligands for Cu^{2+}. Additional coordination sites are occupied by solvent molecules (Appendix A.2.4).

Other hints for the loose coordination of Cu^{2+} by T3b can be found in CW EPR spectra. The four low-field peaks of the 'free' and 'loosely bound' Cu^{2+} species are asymmetrically broadened towards a higher magnetic field (Fig. 4.7). As shown in Appendix A.2.3, this broadening originates from an antiproportional relationship between $g_{\parallel} - g_e$ and $A_{\parallel}$ and a coupled strain of both parameters. Thus, the magnitude of this asymmetric broadening is directly related to ligand field inhomogeneities of the copper complexes. The largest relative broadening is observed for free $CuCl_2$ in the ternary solvent mixture due to the variety of potential ligands (water, ethanol, glycerol, and chloride). In contrast, no comparable broadening is observed for the 'bound' and 'intermediate' species, since the copper ion is located in a well-defined ligand field due to the presence of strong, chelating ligands (Fig. A.11). For both 'loosely bound' species, an intermediate broadening is observed. This is a further indication that the coupled triazole units in T3b do not provide for a coordination sphere as well defined as in T3t. They are rather loosely attached to the metal ion.

Another hint for a loose Cu^{2+} coordination is provided by fluorescence spectroscopy. A strong complexation with all three triazole moieties (as observed in T3t) would force the cholic acid arms together at their tails and may such prohibit the encapsulation of pyrene. This is not observed for T0. Though the hosting characteristics were found to be slightly reduced when Cu^{2+} was added to a solution of T3b prior to pyrene, the guest was not prohibited to enter the molecular pocket (Appendix A.2.2).

From this plentitude of observations, it can unambiguously be inferred that T3b —in contrast to T3t—does not act as a metal chelator. Though the copper ion is still attracted by the triazole moieties, it is rather loosely attached to one or two ligands similar to copper complexed with the cholic acid monomer S2. The different triazole groups, although chemically interconnected, do not cooperatively bind to the metal ion but rather act as single entities.

Interestingly, the relative spectral contributions of the loosely bound Cu^{2+} species match the contributions of the intermediate and bound species at the same molar ratios of $CuCl_2$ and the cholic acid derivative (Tables 4.2, 4.4). With the data at hand, it cannot be conclusively answered if this similarity is a mere coincidence or if it originates from the structural properties of the cholic acid oligomers.

4.3.3 Influence of a Fourth Cholic Acid Arm

While the impact of the relative triazole position in the oligomer was in the focus of the last section, it will now be assessed how the structure and function of the oligomer is affected by the addition of a fourth arm to the cholic acid star. Specifically, it is of interest if the fourth cholic acid arm participates in the cavity formation, as suggested in Scheme 4.2, or if the induced bond strain at the connecting carbon atom prohibits the simultaneous alignment of all four side chains.

Table 4.4 Spectral contribution of different copper species in T3b

Ratio	Relative spectral contribution		
$CuCl_2$: T3b	Free	Loosely bound1	Loosely bound2
1:2	<0.03	0.10 ± 0.05	0.87 ± 0.05
1:1	0.12 ± 0.02	0.26 ± 0.05	0.62 ± 0.05

Table 4.5 Spectral contribution of copper species in mixtures of copper and Q4t/Q8tb

Ratio			Relative spectral contribution	
$CuCl_2$: CA oligomer		Free	Intermediate	Bound
Q4t	1:2	-	0.12 ± 0.05	0.88 ± 0.05
	1:1	0.13 ± 0.02	0.29 ± 0.05	0.58 ± 0.05
Q8tb	1:2	-	0.25 ± 0.05	0.75 ± 0.05
	1:1	<0.02	0.35 ± 0.05	0.63 ± 0.05
	2:1	0.13 ± 0.02	0.38[a]	0.50[a]

[a] Additional occurrence of a fourth spectral component (details see main text, Sect. 4.3.4)

For this study, the tetramer Q4t was considered since its fourth cholic acid chain constitutes the only structural difference in comparison to T3t.

In fact, CW EPR revealed the same spectral components as observed in T3t, i.e. strongly coordinated Cu^{2+} ions that are chelated with two and three triazole moieties (spectra not shown). However, a copper species complexed to four nitrogen atoms was absent. This unambiguously proves that the fourth arm of the tetramer is not involved in the Cu^{2+} complexation. Apparently, it cannot bend down towards the other three cholic acid chains and does not participate in the cavity formation.

The quantitative contributions of the observed spectral components are in good agreement with those observed for T3t (Tables 4.2, 4.5). In a 1:2 mixture of $CuCl_2$ and Q4t, slightly more Cu^{2+} is coordinated to three triazole groups (88% vs. 82% in T3t). With a total of four arms available, the binding to three triazole moieties may be statistically favored in comparison to a total of three arms. In an equimolar mixture, almost identical fractions of intermediate (29%/32%) and bound (58%/58%) species were observed for Q4t and T3t. Here, the statistical advantage of a fourth cholic acid arm could be compensated by a second effect. The availability of more Cu^{2+} ions could force the tetramer to structurally separate into two pairs of cholic acid arms. By this, the tetramer would provide two intermediate binding sites rather than one strong and one weak site.

For both mixtures, no 'loosely bound' copper could be detected. In principle, this species could originate from a copper ion coordinated to the triazole unit of the single remaining cholic acid chain. Rather, all ^{14}N coordinated metal ions experience the strong ligand field of di- and tridentate ligands, which is energetically and entropically favored over the weak coordination of a single triazole ligand.

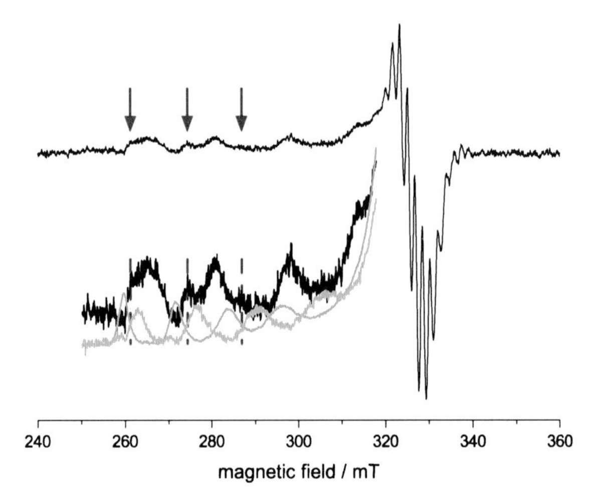

Fig. 4.9 Linear combination $R = \text{Q8tb}(2{:}1) - 0.67\ Q8\text{tb}(1:1) + 0.12\ Q8\text{tb}(1:0)$ of different $CuCl_2$–Q8tb spectra in an attempt to isolate the 'intermediate' copper species. *Red* arrows and dashed lines mark an additional contribution of a fourth species which could not be separated from the 'intermediate' species. The low field region of the spectrum is magnified by a factor of ~15, and compared to the spectra of the 'free' (*green*) and 'loosely bound1' (*turquoise*) species. The fourth spectral component cannot be reproduced with either 'free' or 'loosely bound1' Cu^{2+}, but rather exhibits EPR parameters intermediate between both species

4.3.4 Observation of a Fourth Species in a 2:1 Mixture of $CuCl_2$ and Q8tb

As a cholic acid tetramer with triazole moieties at the head and tail of each chain, Q8tb provides a total of eight ligands for the Cu^{2+} ion. But as already observed for Q4t, up to a molar ratio of 1:1 the metal ion almost exclusively binds to the chelating triazole units near the connection point of the cholic acid arms that provide the strongest ligand field and the most favorable coordination sphere (Table 4.5).

At a copper–oligomer ratio of 2:1, not all Cu^{2+} ions can be hosted in this prime binding spot. A fraction of 'loosely bound' Cu^{2+} is expected in analogy to T3b. In fact, when trying to extract the 'intermediate' species by a linear combination of different CuCl2–Q8tb spectra, a fourth spectral component was observed, which exhibits features of a 'loosely bound' species (Fig. 4.9). Surprisingly, it cannot be assigned to either 'loosely bound1' or 'loosely bound2' copper. Rather, it constitutes a new distinct species, which exhibits EPR parameters $g_{\|} = 2.393 \pm 0.002$ and $A_{\|} = (427 \pm 3)$ MHz, intermediate of those observed for the 'free' and 'loosely bound1' spectral compound. Due to the multiplicity of other spectral contributions and lack of resolution, it was not possible to quantitatively determine all EPR parameters and the relative contribution of this species to the spectrum.

Note that a small amount may also be present in the equimolar mixture of $CuCl_2$ and Q8tb.

As stated in Sect. 4.3.2, the 'loosely bound1' species originates from a Cu^{2+} ion, which is weakly coordinated to the triazole unit of a single cholic acid arm. In comparison, the triazole binding of the species observed here imposes an even smaller ligand field. Hence, the complexation of Cu^{2+} to the chelating triazoles at the head of the cholic acid chain may lead to a structural conformation of its tail that restricts further binding to a second Cu^{2+} ion. Two conclusions can be drawn from this observation. First, the triazole moieties at the tail are not in close proximity to each other since no cooperative binding is observed. Second, they may be shielded from a direct access of the Cu^{2+} ion, so that an efficient coordinative binding to a nitrogen atom of the triazole unit is prohibited. In a first tentative picture, this effect may originate from a partial back folding of the aliphatic part of the cholic acid chain into the hydrophobic cavity.

4.3.5 Summary

This part focused on the influences of structural variations on the spatial arrangement and function of the oligomers. In contrast to T3t, no chelating effect was observed when the triazole group is located at the tail of the cholic acid chains in the oligomers. This resulting weak coordination of the metal ion did not lead to an enhanced fluorescence quenching efficiency. In a cholic acid tetramer, the fourth arm does not participate in the formation of the cavity and rather acts as a single entity. When each arm of an oligomer contains two triazole moieties, the copper complexation at the chain head restricts further binding of a second Cu^{2+} equivalent at the tail of the chain.

4.4 Self-Assembly of Monomeric Cholic Acid Derivatives

In the last part of this chapter, the focus shifts from star-shaped oligomers to a triazole-modified cholic acid monomer. While the cholic acid chains of all oligomers are chemically attached to each other, this covalent link is absent for the monomer. Specifically, the role of Cu^{2+} ions for the self-organization of the single chains is investigated in detail.

4.4.1 Cu^{2+} Coordination by S2

As already mentioned in Sect. 4.3.2, $CuCl_2$–S2 mixtures exhibit the same spectral species as $CuCl_2$–T3b conjugates, proving that the triazole moieties in T3b act as

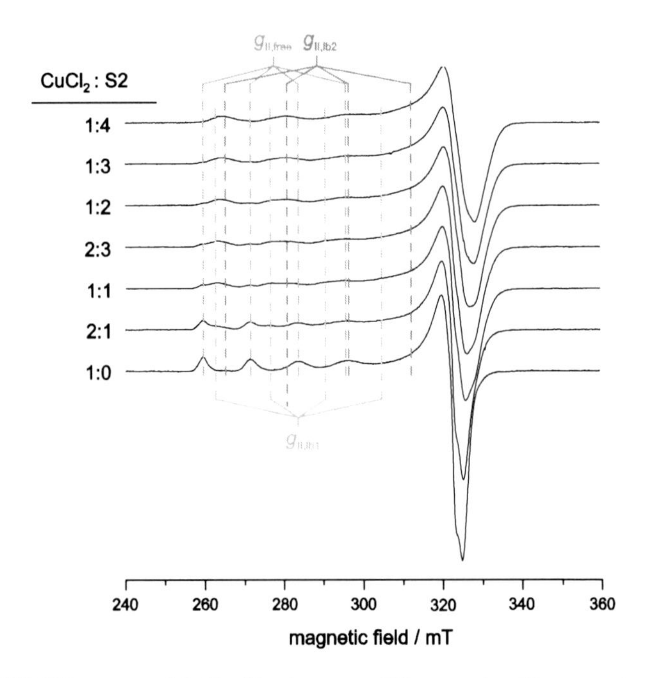

Fig. 4.10 CW EPR spectra of $CuCl_2$–S2 mixtures at different ratios. The spectral contributions of all Cu^{2+} species in the low field region are visualized by vertical dashed lines: free (*green*), loosely bound1 (*turquoise*), loosely bound2 (*ochre*)

single entities. The CW EPR spectra of a variety of $CuCl_2$–S2 mixtures are displayed in Fig. 4.10 and the spectral contributions of the 'free', 'loosely bound1', and 'loosely bound2' species are highlighted by vertical lines.

The relative ratios of the three spectral compounds are quantified and listed in Appendix A.2.5. A graphical illustration is given in Fig. 4.11. The fraction of uncoordinated Cu^{2+} decreases rapidly as more S2 molecules become available. This decrease is fitted with a mono-exponential decay, which reveals with a characteristic ratio χ_0 of 1.19 ± 0.11. At this S2/$CuCl_2$ ratio, the fraction of free Cu^{2+} is decreased to 1/e of its original value. Thus, one di-substituted cholic acid monomer is able to complex one Cu^{2+} ion on average. At $\chi = 4$, only a negligible Cu^{2+} fraction remains uncoordinated. The contribution of the species 'loosely bound1' increases to a maximum value of 48% at an equimolar ratio of $CuCl_2$ and S2. Above this ratio, the relative amount of this species decreases as increasingly more Cu^{2+} ions are coordinated to two cholic acid chains. The fraction of this third species 'loosely bound2' increases linearly with χ. At $\chi = 4$, already over 90% of all Cu^{2+} ions are coordinated to two cholic acid chains.

Despite the fact that the relative fractions of all copper species undergo pronounced changes with respect to the $CuCl_2$–S2 ratio, no such trend can be observed in the HYSCORE data (Fig. 4.12). The recorded spectra exhibit features that bear large similarities to those of $CuCl_2$–T3b mixtures (Fig. 4.8a). Due to an

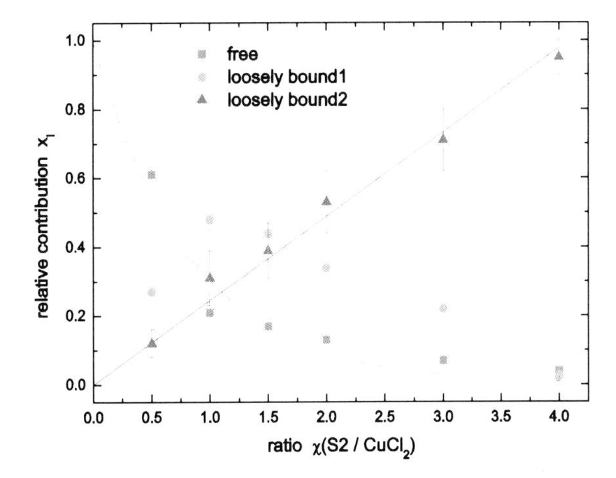

Fig. 4.11 Graphical representation of the relative contributions of all spectral species for different Cu–S2 ratios $\chi = n(S2)/n(CuCl_2)$. The fraction of the free copper species was fitted with an exponentially decaying function $x_{free} = \exp(-\chi/\chi_0)$ with $\chi_0 = 1.19 \pm 0.11$, while fraction 'loosely bound2' could be represented by a first order polynomial $x_{ib2} = m\chi$ with a slope $m = 0.246 \pm 0.009$. The 'loosely bound1' data points were fitted by the residual $x_{ib1} = 1 - x_{ib2} - x_{free}$.

enhanced spectral intensity, more transitions are resolved and a spectrum at the magnetic field position corresponding to $g_{\parallel}$ could be recorded. With spectra at both, $g_{\perp}$ and $g_{\parallel}$ the strongly coupled nitrogen nuclear spin could be studied in more detail. Preliminary simulations suggest hyperfine contributions $A_{\perp} \cong 6$ MHz and $A_{\parallel} \cong 3$ MHz, a nuclear quadrupole frequency $K = 2.3$ MHz, and an anisotropy parameter of $\eta = 0.9$.

4.4.2 Cu^{2+} Mediated Spatial Assembly of Single Cholic Acid Chains

Each cholic acid monomer is substituted by two triazole units, which may exhibit a well-defined intramolecular distance due to the rigidity of the cholic acid building block. If both units functioned as ligands for Cu^{2+} ions, these ions should also show the same distance relationship, which in turn should be accessible by double electron–electron resonance spectroscopy (DEER). This defined distance relation may even be further extended to three Cu^{2+} ions by a non-covalent interconnection of two cholic acid chains by Cu^{2+} species 'loosely bound2'.

However, time domain signals of $CuCl_2$–S2 conjugates in a ternary solvent mixture of ethanol, water, and glycerol follow a mono-exponential decay, which suggests a homogeneous spatial distribution of the metal ions in three dimensions. In contrast, the time trace differs significantly from an exponential decay, when the

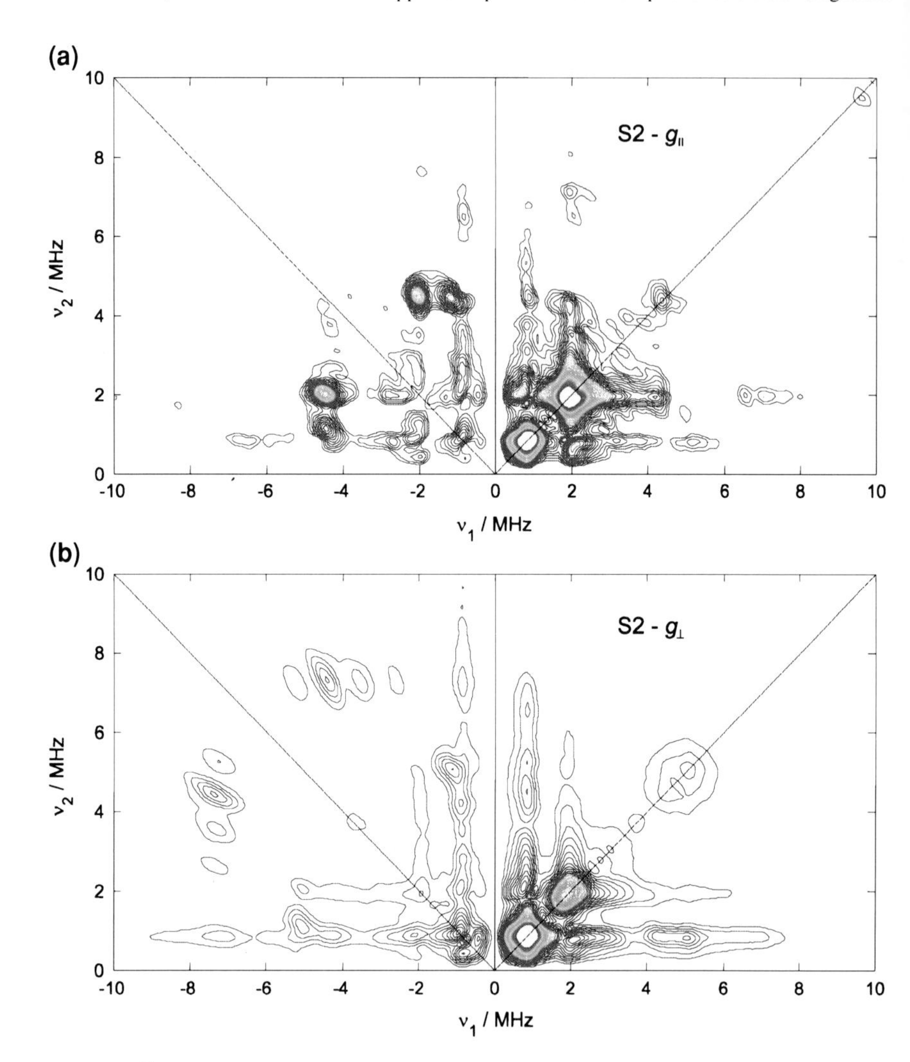

Fig. 4.12 ^{14}N region of the HYSCORE spectra of a 2:3 mixture of $CuCl_2$ and S2. The spectra were recorded at the EPR spectral maximum (corresponding to $g_{\perp}$, (**b**) and at a magnetic field position corresponding $g_{\parallel}$ (**a**). The spectra were recorded at 10 K, τ was set to 132 ns

solvent composition is changed to pure ethanol. The background corrected DEER time domain signal and the extracted distance distribution are shown in Fig. 4.13. Given the large ^{1}H nuclear modulations, a Tikhonov regularization parameter of 1,000 could be deemed reasonable, which gives rise to a broad distance distribution with two major contributions around 1.6 and 3.0 nm. By lowering the parameter to 10, the single contributions to these broad peaks are revealed. They are centered at 1.6, 2.1, 2.7, 3.2, and 3.9 nm.

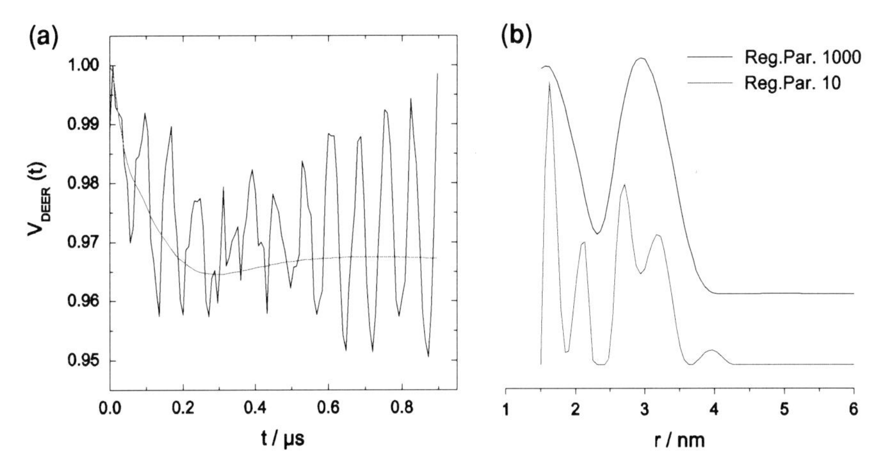

Fig. 4.13 **a** Background corrected DEER time domain signal (*black*) and fit (*red*) of a 2:1 mixture of $CuCl_2$ and S2. The pump frequency was set to the maximum of the absorption spectrum, the observer pulses were set to a 65 MHz higher frequency. **b** Extracted distance distributions by Tikhonov regularization with regularization parameters of 1,000 (*black*, corresponding to the corner of the L curve) and 10 (*blue*)

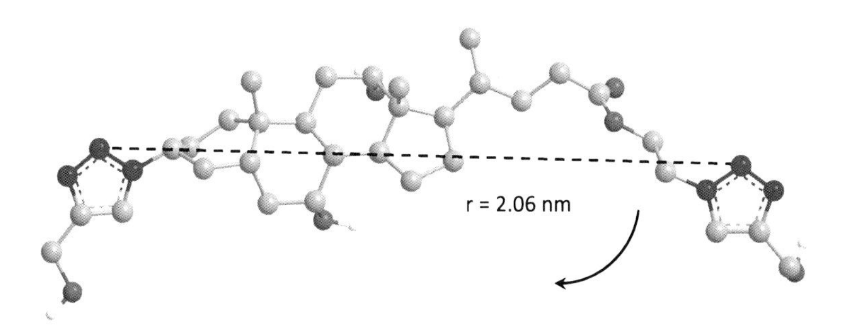

Scheme 4.3 Force field optimized structure (MMFF94, Chem3D Ultra) of S2. A distance of 2.06 nm is obtained between the central nitrogen atoms of the triazole units. The arrow indicates a potential backfolding of the aliphatic chain, which could lead to a significantly decreased distance between the triazole groups

Force field simulations on S2 suggest a distance of 2.06 nm between the two central nitrogen atoms of the triazole units, which is in remarkable agreement with the observed distance peak at 2.1 nm (Scheme 4.3). One should note that the optimized molecular structure exhibits a rather extended conformation. A potential backfolding of the flexible aliphatic chain would significantly decrease the distance between the two triazole groups, which could well be in the range of the lowest observed distance of 1.6 nm.

Two counteracting effects limit the validity of DEER retrieved distances around 1.5 nm. On the one hand, this distance is significantly overestimated due to residual ESEEM modulations by protons. In the case at hand, this contribution is

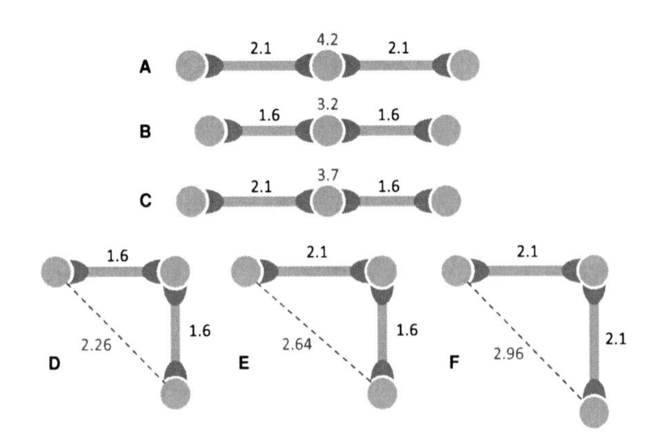

Scheme 4.4 Expected distances between Cu^{2+} ions (*green*) of Cu(II)-linked cholic acid dimers. Two distinct intramolecular triazole distances of 2.1 and 1.6 nm for extended and backfolded cholic acid chains were assumed. The cholic acid chains were treated as rigid rods, and an octahedral coordination of the Cu^{2+} ion was assumed. All distance values are given in nm

very pronounced due to the low modulation depth of $\lambda = 0.032$ (cf. Fig. 4.13a). The frequency of this modulation, the Larmor frequency of protons $\nu(^{1}H) = 14.2$MHz, matches the dipolar electron–electron coupling at a distance of ~ 1.5 nm. On the other hand, distances smaller than 2 nm are underestimated due to the limited excitation bandwidth of the pump pulse. Since both effects counteract each other, they are not accounted for in the analysis of the data. Hence, the observed distance around 1.6 nm needs to be treated with care. Nevertheless, from the force-field calculations one can safely assume that the cholic acid oligomers can exhibit two well-defined conformations. An extended conformation causes a distance of 2.1 nm between two Cu^{2+} ions while a backfolded conformation leads to a distance of 1.6 nm. This is needed for the following considerations.

In a simplified picture, two cholic acid chains that are linked by an octahedrally coordinated Cu^{2+} ion can be described as a linear or vertical connection of two rigid rods (Scheme 4.4). In this case, the distances between the peripheral Cu^{2+} ions are described the sum of the single rod distances or the hypotenuse of a right-angled triangle, as depicted in red. By combinations of the extended and the backfolded conformation, six different potential distances are obtained. These combination distances of 2.26, 2.64, 2.96, 3.2, 3.7, and 4.2 nm match remarkably well with the observed DEER distance distribution despite the fact that the single cholic acid chains exhibit a large conformational flexibility due to their significant aliphatic parts. This conformational freedom leads to a broad distance distribution instead of well-defined peaks. Note that with an evolution time of 0.9 μs, distances above 4.2 nm cannot be observed and distances above 3.6 nm could be truncated.

It is important to note that the analysis applied here does not consider the orientational selection of the excited spin packets in contrast to the data evaluation in Sect. 3.4. Further, the low modulation depth and the bad SNR only allowed for a DEER measurement at the spectral maximum, which contains orientations predominantly along the direction perpendicular to the unique octahedral axis of the copper complex. Yet, pronounced orientational artefacts in the DEER data of the coordinated Cu^{2+} ions are weak and can generally neglected as the binding sites are not rigidly oriented but have large orientational freedom. Multispin effects due to the excitation of several pumped spins should be negligible due to the broad spectral distribution and low inversion probability, which manifests in the low modulation depth.

With all care that needs to be applied due to possible pitfalls of the DEER experiment, it can be stated that (at least) two cholic acid units are non-covalently interconnected by Cu^{2+} ions. Despite the weak Cu–triazole coordination and the conformational flexibility of the cholic acid chains, this self-assembly manifests in a well-defined spatial distribution of the complexed metal ions.

4.5 Conclusions

In this chapter, the functional structure of cholic acid based oligomers was characterized by EPR and fluorescence spectroscopy. These oligomers show a complex molecular self-organization due to the amphiphilicity of their building blocks and can coordinate metal ions due to incorporated 1,2,3-triazole units. It was found that a preordered arrangement of the ligands is crucial for a strong complexation of metal ions. In case of T3t, this chelate-like binding of metal ions leads to a close contact between metal and host. If pyrene is hosted in the molecular pocket, a tremendous enhancement of the fluorescence quenching is observed, which can be utilized for the sensing of heavy metal ions down to the ppb range. The oligomer may also act as catalytic center since many reactions are accelerated in the presence of transition metals. In this context, it would constitute a synthetic analogue to the active site of certain enzymes.

Without a preordering, every triazole unit acts as a single entity. In the absence of a chelating effect, both, coordination strength and number of coordinated triazole ligands are decreased. Due to this weak ligation, the metal ion does not show a strong impact on the fluorescence quenching efficiency anymore. Yet, these decreased complexation properties are still sufficient to induce a Cu^{2+} mediated self-assembly of di-substituted cholic acid monomers. Under optimized conditions, this self-assembly could lead to the formation of a non-covalent supramolecular network.

From an EPR spectroscopic point of view, Cu^{2+} was subjected to a large variety of triazole coordinations. Six different Cu(II)–triazole conjugates were indentified and characterized. The observed complexes cover a wide range of binding strengths and contain one to three coordinated triazole ligands. Thus, they may serve as a reference for further EPR studies on Cu(II)–triazole interactions. This method may be increasingly used as a means of structural characterization as on the one hand, EPR data of coordinated Cu^{2+} ions deliver a plentitude of information about the system and on the other hand, triazole units are present in many synthetic molecules due to the widely used ‘click’ reaction.

4.6 Materials and Methods

Materials. Pyrene (Sigma-Aldrich, 99%), anhydrous $CuCl_2$ (purum, Fluka), ethanol (HPLC grade, Sigma-Aldrich), and glycerol (87 wt%, Fluka) were used as received. Distilled water was further purified by a MilliQ System (Millipore) to achieve a resistivity of 18.2 MΩ cm. All cholic acid derivatives used in this study were synthesized by Juntao Luo and Jiawei Zhang. The synthetic procedures are reported elsewhere [23, 57].

Fluorescence Studies. All fluorescence experiments were conducted by Jiawei Zhang. A stock solution of pyrene (20 μM in methanol) was added to a solution of the cholic acid oligomers, dissolved in MilliQ water (Sect. 4.2) or in a mixture of water and methanol (Sect. 4.3). For the fluorescence quenching studies, stock solutions of metal ions (5 mM or 1 M in water) were added to obtain the desired concentrations of metal ions. To study if Cu^{2+} prevents pyrene from entering the hydrophobic cavity of T3b, 1 eq. Cu^{2+} was admixed to the T3b solution before pyrene was added (Appendix A.2.2). The final concentration of pyrene was adjusted to 0.1 or 0.2 μM, the ratio of water and ethanol in the final mixture was 99:1 (v/v, Sect. 4.2) or 95:5 (v/v, Sect. 4.3). After degassing the samples, steady-state fluorescence spectra were recorded at room temperature on a Varian fluorescence spectrophotometer equipped with Xe-900 lamp with an excitation wavelength of 335 nm. The excitation and emission slit widths were 10 and 2.5 nm, respectively.

Sample Preparation for EPR Measurements. 5 mM $CuCl_2$ in EtOH was complexed with cholic acid derivatives in different molar ratios. Ethanol, water, and glycerol were added to achieve a solvent composition of 9:1:2 (v/v). The ternary solvent mixture increases the solubility of the cholic acid derivatives and prevents crystallization upon freezing. The final concentration of Cu^{2+} was adjusted to 2 mM. The mixtures were placed in EPR quartz tubes (i.d. 3 mm) and shock-frozen in $N_2(l)$ cooled isopentane to yield fully transparent glasses.

CW EPR Measurements and Analysis. Continuous wave (CW) EPR spectra of the $CuCl_2$–oligomer mixtures were recorded on a Miniscope MS200 benchtop spectrometer (Magnettech, Berlin, Germany) working at X-band (~9.4 GHz) with a modulation amplitude of 0.2 mT, a sweep width of 140 mT, a sweep time of 120 s, and a microwave power of 6.2 mW. A manganese standard reference (Mn^{2+} in ZnS, Magnettech) was used to calibrate the magnetic field. All spectra were recorded at 78 K. The spectra were background-corrected (subtraction of a recorded baseline, first and zeroth order baseline correction) and normalized to their double integral (with exception of the spectra in Fig. 4.3, which were normalized to their maximum value).

The contributing Cu^{2+} species were isolated by linear combinations of spectra with different $CuCl_2$–oligomer mixtures. They were simulated with a home-written Matlab program, assuming collinear uni-axial **g** and **A** tensors and a natural isotopic composition of Cu and utilizing the slow motion routine of the Easyspin software package for EPR (see Sect. 2.7) [62]. The spectra were then reproduced by linear combination of the normalized (double integral) spectral contribution (Sects. 4.3, 4.4) or their simulations (Sect. 4.2) to obtain the relative fractions of the spectral species.

Pulse EPR Measurements. All pulse EPR experiments were performed at X-band frequencies (9.2–9.4 GHz) with a Bruker Elexsys 580 spectrometer equipped with a Bruker Flexline split-ring resonator ER4118X_MS3. The temperature was set to 10 K by cooling with a closed cycle cryostat (ARS AF204, customized for pulse EPR, ARS, Macungie, PA, USA). First, an ESE detected spectrum was recorded with the primary echo sequence $\pi/2 - \tau - \pi - \tau -$ echo. τ was set to 200 ns (protonated solvents) or 400 ns (deuterated solvents). 3-Pulse ESEEM and HYSCORE spectra were recorded at B_{max}, $B_{max} - 25$ mT and, where possible, at the spectral position corresponding to $g_{\|}$.

3-Pulse Electron Spin Echo Envelope Modulation (ESEEM). 3-Pulse ESEEM spectra were recorded using the stimulated echo sequence $\pi/2 - \tau - \pi/2 - T - \pi/2 - \tau -$ echo [58], with pulse lengths of $t_p = 16$ ns. The time intervals of t_1 and t_2 were varied from 300 to 8,492 ns in steps of 16 ns. A four-step phase cycle was used to eliminate unwanted echoes. The time traces were baseline corrected with a third-order polynomial, apodized with a Gauss window, and zero-filled. After two-dimensional Fourier transformation, the absolute value spectra were calculated. To eliminate blind spots, spectra with τ values of 132, 164, and 200 ns were added up.

HYSCORE Spectra and Simulation. HYSCORE spectra were measured with the pulse sequence $\pi/2 - \tau - \pi/2 - t_1 - \pi - t_2 - \pi/2 - \tau -$ echo [59]. The length of all pulses was set to $t_p = 16$ ns. The time intervals of t_1 and t_2 were varied from 300 to 4,396 ns in steps of 16 ns. An eight-step phase cycle was used to eliminate unwanted echoes. The time traces of the HYSCORE spectra were baseline corrected with a third-order polynomial, apodized with a Gauss window, and zero-filled. After two-dimensional Fourier transformation, the absolute value spectra were calculated. To account for blindspot artifacts, HYSCORE spectra were recorded with two τ values (132 and 164 ns) and analyzed separately. The spectra were simulated with a Matlab program written by Madi [63].

DEER Distance Measurements and Analysis. Dipolar time evolution data were obtained using the four-pulse DEER experiment [64–66], as described in detail in Sect. 3.6. The presented spectrum in Sect. 4.4 was obtained from a 2:1 $CuCl_2$-S2 mixture dissolved in pure ethanol. The raw time domain DEER data was processed with the program package DeerAnalysis2008 [67]. Intermolecular contributions were removed by division by a mono-exponential decay. The resulting time traces were normalized to $t = 0$. Distance distributions were obtained by Tikhonov regularization with varying regularization parameters.

References

1. Mukhopadhyay S, Maitra U (2004) Curr Sci 87:1666–1683
2. Benrebouh A, Avoce D, Zhu XX (2001) Polymer 42:4031–4038
3. Fu H-L, Cheng S-X, Zhang X-Z, Zhuo R-X (2007) J Controlled Release 124:181–188

4. Gauthier M, Simard P, Zhang Z, Zhu XX (2007) J R Soc Interface 4:1145–1150
5. Zhang J, Zhu XX (2009) Sci Chin Ser B: Chem 52:849–861
6. Zhu XX, Nichifor M (2002) Acc Chem Res 35:539–546
7. Vijayalakshmi N, Maitra U (2006) J Org Chem 71:768–774
8. Zhong ZQ, Zhao Y (2007) Org Lett 9:2891–2894
9. Whitmarsh SD, Redmond AP Sgarlata V, Davis AP (2008) Chem Commun 3669–3671
10. Janout V, Lanier M, Regen SL (1996) J Am Chem Soc 118:1573–1574
11. Janout V, Zhang LH, Staina IV, Di Giorgio C, Regen SL (2001) J Am Chem Soc 123: 5401–5406
12. Janout V, Jing BW, Regen SL (2002) Bioconjugate Chem 13:351–356
13. Janout V, Jing BW, Staina IV, Regen SL (2003) J Am Chem Soc 125:4436–4437
14. Janout V, Regen SL (2005) J Am Chem Soc 127:22–23
15. Janout V, Jing BW, Regen SL (2005) J Am Chem Soc 127:15862–15870
16. Janout V, Regen SL (2009) Bioconjugate Chem 20:183–192
17. Zhao Y, Ryu EH (2005) J Org Chem 70:7585–7591
18. Ryu EH, Yan J, Zhong Z, Zhao Y (2006) J Org Chem 71:7205–7213
19. Ryu EH, Zhao Y (2004) Org Lett 6:3187–3189
20. Ryu EH, Zhao Y (2006) J Org Chem 71:9491–9494
21. Mukhopadhyay S, Maitra U, Ira, Krishnamoorthy G, Schmidt J, Talmon Y (2004) J Am Chem Soc 126:15905–15914
22. Maitra U, Mukhopadhyay S, Sarkar A, Rao P, Indi SS (2001) Angew Chem Int Ed 40: 2281–2283
23. Luo J, Chen Y, Zhu XX (2007) Synlett 2201–2204
24. Luo J, Chen Y, Zhu XX (2009) Langmuir 25:10913–10917
25. Zhang J, Luo J, Zhu XX, Junk MJN, Hinderberger D (2010) Langmuir 26:2958–2962
26. Chen Y, Luo J, Zhu XX (2008) J Phys Chem B 112:3402–3409
27. Tornoe CW, Christensen C, Meldal M (2002) J Org Chem 67:3057–3064
28. Rostovtsev VV, Green LG, Fokin VV, Sharpless KB (2002) Angew Chem Int Ed 41: 2596–2599
29. Bock VD, Hiemstra H, van Maarseveen JH (2005) Eur J Org Chem 1:51–68
30. Kalyanasundaram K, Thomas JK (1977) J Am Chem Soc 99:2039–2044
31. Lakowicz JR (1999) Principles of fluorescence spectroscopy. Springer, Berlin
32. Assour JM (1965) J Chem Phys 43:2477–2489
33. Stankowski J, Wieckowski A, Hedewy S (1974) J Magn Reson 15:498–509
34. Bohandy J, Kim BF (1977) J Magn Reson 26:341–349
35. Cunningham KL, McNett KM, Pierce RA, Davis KA, Harris HH, Falck DM, McMillin DR (1997) Inorg Chem 36:608–613
36. van Koningsbruggen PJ, van Hal JW, de Graaff RAG, Haasnoot JG, Reedijk J (1993) J Chem Soc Dalton Trans 2163–2167
37. van Koningsbruggen PJ, Gatteschi D, Degraaff RAG, Haasnoot JG, Reedijk J, Zanchini C (1995) Inorg Chem 34:5175–5182
38. Lu J, Bender CJ, McCracken J, Peisach J, Severns JC, McMillin DR (1992) Biochemistry 31:6265–6272
39. Place C, Zimmermann J-L, Mulliez E, Guillot G, Bois C, Chottard J-C (1998) Inorg Chem 37:4030–4039
40. Burns CS, Aronoff-Spencer E, Dunham CM, Lario P, Avdievich NI, Antholine WE, Olmstead MM, Vrielink A, Gerfen GJ, Peisach J, Scott WG, Millhauser GL (2002) Biochemistry 41:3991–4001
41. Benesi HA, Hildebrand JH (1949) J Am Chem Soc 71:2703–2707
42. Chang K-C, Su I-H, Senthilvelan A, Chung W-S (2007) Org Lett 9:3363–3366
43. Lehn J-M (1995) Supramolecular Chemistry: concepts and perspectives. Wiley, New York
44. Qi X, Jun EJ, Xu L, Kim S-J, Hong JSJ, Yoon YJ, Yoon J (2006) J Org Chem 71:2881–2884
45. Xiang Y, Tong A, Jin P, Ju Y (2006) Org Lett 8:2863–2866
46. Park SM, Kim MH, Choe J-I, No KT, Chang S-K (2007) J Org Chem 72:3550–3553

47. Huang S, Clark RJ, Zhu L (2007) Org Lett 9:4999–5002
48. Moon SY, Youn NJ, Park SM, Chang SK (2005) J Org Chem 70:2394–2397
49. Zhao Y, Zhong Z (2006) Org Lett 8:4715–4717
50. Hung H-C, Cheng C-W, Ho I-T, Chung W-S (2009) Tetrahedron Lett 50:302–305
51. Zhu L, dos Santos O, Koo CW, Rybstein M, Pape L, Canary JW (2003) Inorg Chem 42:7912–7920
52. Richeter S, Rebek J (2004) J Am Chem Soc 126:16280–16281
53. Parkin G (2004) Chem Rev 104:699–767
54. Schweiger A, Jeschke G (2001) Principles of pulse electron paramagnetic resonance. Oxford University Press, Oxford
55. Weil JA, Bolton JR, Wertz JE (1994) Electron paramagnetic resonance: elementary theory and practical applications. Wiley, New York
56. Iwaizumi M, Ohba Y, Iida H, Hirayama M (1984) Inorg Chim Acta 82:47–52
57. Zhang J (2010) 基于胆酸的分子篮的合成与性质表征 (The synthesis and properties of molecular pockets based on cholic acid). Doctoral Dissertation, Nankai University, Tianjin
58. Schweiger A (1991) Angew Chem Int Ed 30:265–292
59. H—fer P, Grupp A, Nebenführ H, Mehring M (1986) Chem Phys Lett 132:279–282
60. Jiang F, McCracken J, Peisach J (1990) J Am Chem Soc 112:9035–9044
61. Slutter CE, Gromov I, Epel B, Pecht I, Richards JH, Goldfarb D (2001) J Am Chem Soc 123:5325–5336
62. Stoll S, Schweiger A (2006) J Magn Reson 178:42–55
63. Madi ZL, Van Doorslaer S, Schweiger A (2002) J Magn Reson 154:181–191
64. Milov AD, Salikhov KM, Shirov MD (1981) Fiz Tverd Tela 23:975–982
65. Pannier M, Veit S, Godt A, Jeschke G, Spiess HW (2000) J Magn Reson 142:331–340
66. Jeschke G, Pannier M, Spiess HW (2000) Double electron–electron resonance. In: Berliner LJ, Eaton GR, Eaton SS (eds) Biological magnetic resonance, Distance measurements in biological systems by EPR, vol 19. Kluwer Academic, New York
67. Jeschke G, Chechik V, Ionita P, Godt A, Zimmermann H, Banham J, Timmel CR, Hilger D, Jung H (2006) Appl Magn Reson 30:473–498
68. Zhang J, Junk MJN, Luo J, Hinderberger D, Zhu XX (2010) Langmuir 26:13415–13421

Chapter 5
Nano-Inhomogeneities in Structure and Reactivity of Thermoresponsive Hydrogels

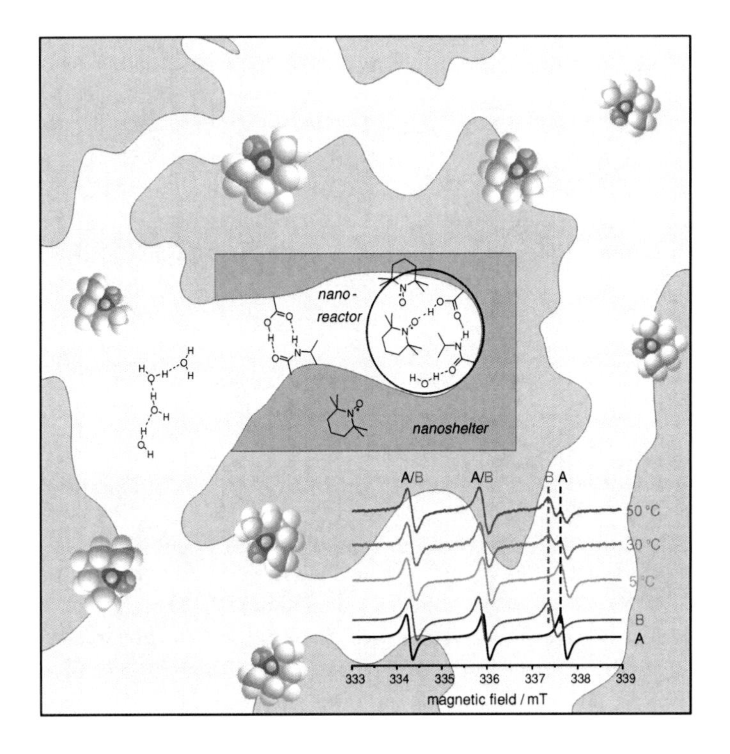

The dynamic and chemical behavior of solute molecules inside thermoresponsive hydrogels based on *N*-isopropylacrylamide is studied by continuous wave EPR spectroscopy. Via addition of paramagnetic tracer molecules, so-called spin probes, the thermally induced collapse is monitored on the molecular scale and is found to proceed over a substantially broader temperature range than indicated by the sharp macroscopic volume transition. The sampling of hydrophilic and

M. J. N. Junk, *Assessing the Functional Structure of Molecular Transporters by EPR Spectroscopy*, Springer Theses, DOI: 10.1007/978-3-642-25135-1_5,

hydrophobic environments suggests a discontinuous collapse mechanism with a coexistence of collapsed and expanded network regions. These static structural inhomogeneities on the nanoscale also lead to an inhomogeneity in chemical reactivity. The hydrophilic regions form nanoreactors, which strongly accelerate chemical reactions while the hydrophobic regions act as nanoshelters, in which enclosed spin probes are protected from the decay. The results show that the copolymer system displays a remarkably complex behavior that mimics spatial and chemical inhomogeneities observed in functional biopolymers such as enzymes.

5.1 Introduction

Responsive hydrogels constitute an interesting class of polymeric materials. While ordinary hydrogels are hydrophilic, crosslinked polymer networks and thus swell in water at all conditions, responsive hydrogels exhibit sufficient hydrophilicity only in a certain range of conditions. These conditions can be changed by variation of external parameters such as temperature, pH, pressure etc. [1, 2]. Thus, a reversible phase separation can be induced. The incorporated water is driven out of the polymeric network and the system collapses macroscopically. This switchability has attracted much attention in the scientific community due to the large variety of possible applications in biology, medicine, pharmacy, and as sensors and actuators [3–8].

A well-studied representative of thermoresponsive polymeric materials is poly(*N*-isopropylacrylamide) (pNiPAAm) in various modifications such as networks or surface grafted-brushes [9]. The homopolymer exhibits a lower critical solution temperature (LCST) close to body temperature (32 °C), above which the polymer aggregates and expels bound water while at $T <$ LCST, water is a good solvent for the polymer [9–11]. By temperature increase above the LCST the polymer and the water molecules phase separate, leading to a macroscopic network collapse of the hydrogel with incorporated water being released. This phase separation originates from a shift of the balance between the hydrophilic and hydrophobic interactions with respect to temperature [9, 12, 13], or in thermodynamic terms, between the mixing free energy and the rubber elasticity free energy, and in the presence of ionizable groups the osmotic effect of all charged species (see Ref. [17–21] in citation [14]).

Though responsive hydrogels and especially pNiPAAm have been studied extensively in the literature, the bulk of the scientific research dealt with the dependence of the chemical structure on the macroscopic swelling behavior. Considerably less publications have looked at the structural morphology of those hydrogels, despite the fact that microstructure and inhomogeneities on the submicrometer length scale have a significant influence on the swelling and collapse behavior [15, 16].

In this chapter, an approach is presented to obtain information of the system from a guest molecule's point of view. The specific characteristics of responsive hydrogels

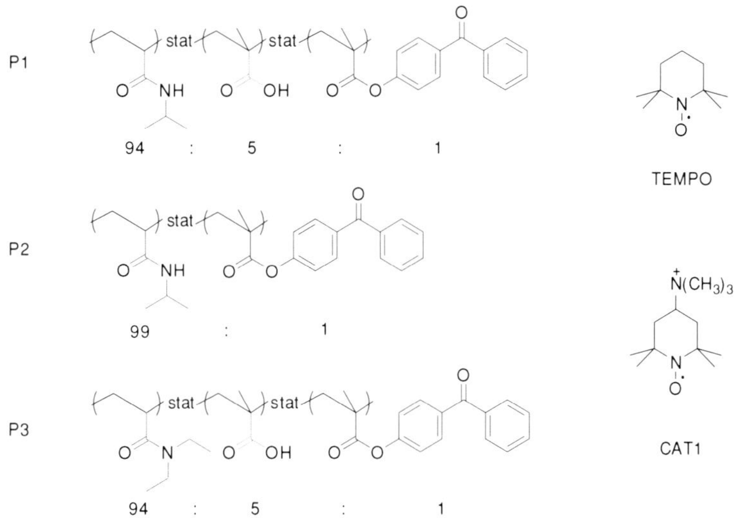

P1

stat stat

O NH O OH O O

94 : 5 : 1

TEMPO

P2

stat

O NH O O

99 : 1

$\overset{+}{N}(CH_3)_3$

P3

stat stat

O N O OH O O

94 : 5 : 1

CAT1

Scheme 5.1 Chemical structures of the non-crosslinked polymeric systems (*left*) and spin probes (*right*). The numbers denote the ratios of the monomers in the polymerization feed

are probed via addition of nitroxides as paramagnetic tracer molecules. Since these spin probes show a pronounced sensitivity for their immediate environment, they provide a picture of the collapse process at a molecular level [17–21].

The material investigated here consists of a responsive hydrogel system that can be spin-coated as thin photocrosslinkable films with a thickness down to a few nanometers [22, 23]. In the terpolymer denoted P1, *N*-isopropylacrylamide (NiPAAm) provides the responsive behavior. Methacrylic acid (MAA) as an ionic component increases the hydrophilicity and prevents the skin effect [9, 24]. The third monomer 4-methacryloyloxybenzophenone (MABP) serves as photocrosslinkable unit [25].

This basic chemical composition of the terpolymer was varied. MAA was excluded to yield the photocrosslinked pNiPAAm copolymer P2 that does not contain any carboxylic acid groups. To understand the specific role of the *N*-isopropyl groups in the NiPAAm units in more detail, these moieties were substituted in P3 by *N*,*N*-diethylacrylamide (DEAAm). This acrylamide derivative provides a comparable collapse behavior of the gels with a LCST of 26–35 °C [26]. The major difference between NiPAAm and DEAAm is the presence of an amide proton in NiPAAm, which may allow hydrogen bond formation with the carbonyl oxygen and water. Chemical structures of all polymers and spin probes are displayed in Scheme 5.1.

This chapter is organized as follows: First, the experimental results obtained with the spin probe approach are described concisely. This is followed by a discussion in light of the hydrogel collapse as seen by probe molecules on the basis of pure *physical* interactions. Subsequently, a peculiar consequence of the observed inhomogeneities on the *chemical* decomposition of the spin probe is presented. The studied hydrogels simultaneously act as nanoshelters, shielding guest molecules from chemical reagents, and as nanoreactors in which chemical reactions of the guest molecules are significantly accelerated in comparison to free solution.

5.2 Results

5.2.1 The Temperature Induced Hydrogel Collapse as Seen by Probe Molecules

As described in Sects. 2.6.3 and 2.6.4, the characteristic three-line pattern of nitroxides in fluid solution is very sensitive towards changes in the rotational dynamics as well as changes in the environment (solvent polarity, hydrophilicity/hydrophobicity etc.). In other words, the radical molecule sensitively probes its environment by *physical* interactions. The presence of local dipoles and the ability of the solvent to form hydrogen bonds leads to a redistribution of electron density in the spin probe, which affects the shape and hyperfine splitting of the EPR signal. While the peak separation is mainly sensitive to the polarity of the environment, the linewidths and relative line amplitudes are governed by the rotational tumbling of the spin probe.

Selected temperature-dependent spectra of TEMPO in hydrogel P1 are displayed in Fig. 5.1. At 5 °C, the observed isotropic hyperfine coupling constant equals that of TEMPO in pure water (data not shown). At this temperature, water is a good solvent for the polymer, leading to strong polymer hydration. The similar coupling constants suggest that at 5 °C the TEMPO molecules are located in a hydrophilic environment and that a swollen hydrogel with large amounts of incorporated water resembles bulk water on EPR length and time scales. However, the rotational correlation time for TEMPO in the swollen hydrogel is slightly larger than in pure water ($\tau_c = (1.8 \pm 0.4) \times 10^{-11}$ s compared to $\tau_c \leq 1.10^{-11}$ s). This is indicative of a reduced rotational mobility of the spin probe due to sterical confinement in the polymer network. A similar effect of reduced mobility in the hydrogel was recently observed for translational diffusion by means of fluorescence correlation spectroscopy [27]. In addition, thin-film photon correlation spectroscopy was employed to determine the dynamic mesh size of the network of 1–5 nm that correlates to such structural inhomogeneities [28].

As the temperature increases, but still well below the critical temperature T_C, the high field line splits into two well defined components, denoted A and B. While the B peak in the high field region grows larger with increasing temperature, the intensity of A decreases at the same ratio. In fact, this second peak B stems from a second type of TEMPO species with an apparent hyperfine splitting that is 4 MHz smaller than that for species A. It can be assigned to TEMPO in collapsed network regions. Since changes of the hyperfine coupling are accompanied by

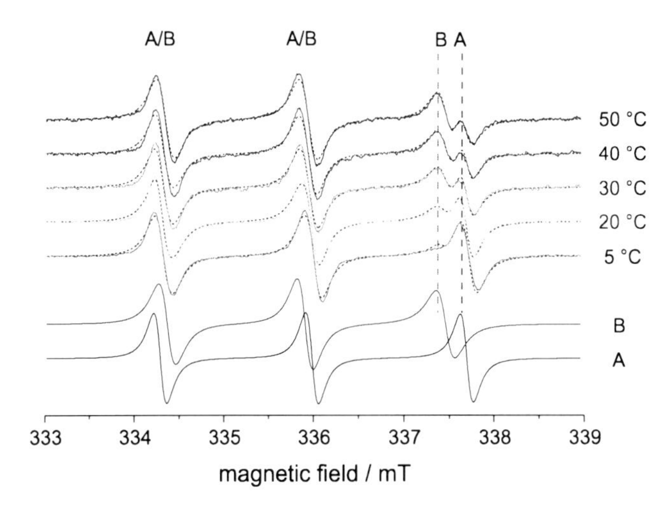

Fig. 5.1 Spectra of a 0.2 mM TEMPO aqueous solution in 120 min crosslinked hydrogel P1 at various temperatures with underlayed intermediate motion simulations (*dotted black lines*). The simulations are based on the assumption of two coexisting spin probe species, denoted A and B, which are shown separately in the graph. Reprinted with permission from [44]. Copyright 2008 WILEY–VCH Verlag GmbH & Co. KGaA, Weinheim

changes of the isotropic g-value, the $m_{\mathrm{I}} = -1$ manifold is most strongly affected. The contributions of A and B to the other manifolds are not resolved.

The magnitude of the hyperfine splitting of nitroxide spin probes is proportional to the polarity and hydrophilicity of the immediate molecular surrounding [29–31]. The significantly reduced hyperfine splitting constant of species B ($a_{\mathrm{B}} = 43.2$ MHz as compared to $a_{\mathrm{A}} = 47.3$ MHz) [32] suggests that with increasing amounts of expelled water more and more polymer chains remain in the direct surrounding of the spin probe, which provide a much more hydrophobic environment for species B than water. This species rather senses a local polarity comparable to that of chloroform or *tert*-butylalcohol [30]. This observation is in good agreement with nuclear magnetic resonance (NMR) results by Kariyo et al. [16]. These authors used the chemical shift difference of ^{129}Xe before and after the collapse to show the existence of hydrophilic and hydrophobic regions in a pNiPAAm-based hydrogel.

Additionally, the rotational motion of spin probes in the two different environments is substantially altered. While the spin probes in swollen meshes can rotate rather freely, the motion in collapsed meshes is strongly restricted due to the largely increased confinement of the polymer network. In fact, the rotational correlation time for species B is increased by a factor of 10 to $\tau_c = (2.0 \pm 0.2) \times 10^{-10}$ s.

The simultaneous observation of two distinct spectral components implies a co-existence of hydrophilic and hydrophobic microenvironments. It is worthwhile to note that the hyperfine coupling constants remain mainly unaffected by temperature variations. This suggests that the two spin probe species do not exchange on the EPR time scale of several ns, i.e. the observed inhomogeneities are static. The small decrease of a_{A} by 0.46 MHz can be explained by a decrease of the

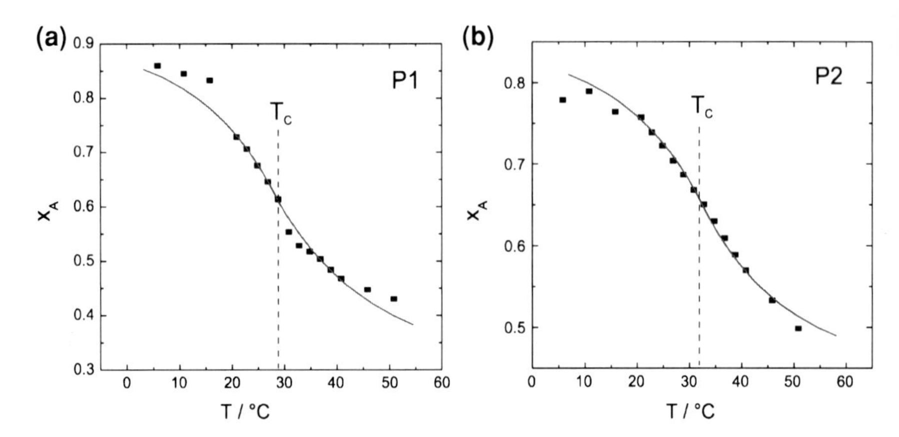

Fig. 5.2 Collapse transition curves of hydrogels **a** P1 and **b** P2. The relative ratios x_A are obtained by spectral simulations and fitted by $x_A(T) = A[1 - 0.5\,\exp(-B(T - T_C))]\,\forall T < T_C$ and $x_A(T) = 0.5A\,\exp(B(T - T_C))\,\forall T < T_C$. The drop between the third and fourth data point in (**a**) is due to a significantly larger waiting time between the two measurements (1 h), which does not change the collapse curve extension and the extracted deflection point. The waiting time between recording all other data points was below 5 min. Reprinted with permission from [44]. Copyright 2008 WILEY-VCH Verlag GmbH & Co. KGaA, Weinheim

dielectric constant of water with increasing temperature [33]. Nevertheless, the decrease is more pronounced than that of free TEMPO in an aqueous solution (−0.35 MHz). This may be caused by an increasing influence of the polymer chains on the spin probes, as the swollen cavities are forced to shrink due to the collapse of surrounding regions. On the other hand, the hyperfine coupling of TEMPO in collapsed regions a_B does not change at all with respect to temperature, as collapsed regions contain much less water and hardly experience a temperature-dependent decrease of the dielectric constant.

The relative contributions of spin probes A and B change with respect to temperature. This ratio is a direct measure of the fraction of swollen and collapsed network regions and thus provides information about the collapse of the hydrogel, as seen by an incorporated molecule. TEMPO as an amphiphilic molecule (with a hydrophilic N–O moiety and the rather hydrophobic hydrocarbon ring) partly stays in the collapsed hydrophobic polymer regions while a certain fraction is expelled into the free water phase in the course of the microphase separation. The ratio A/B in macroscopically fully collapsed gels is thus governed by the volumetric polymer fraction in the sample.

In Fig. 5.2, the fraction x_A of species A in P1 and P2 is plotted against the sample temperature. The fractions were quantitatively accessed by simulating the experimental spectra as a combination of simulations S_{tot} with contributions from species A (S_A, hydrophilic environment) and species B (S_B, hydrophobic environment) according to

$$S_{tot} = x_A S_A + (1 - x_A) S_B. \tag{5.1}$$

The fraction x_A of TEMPO in swollen regions decreases continuously as the temperature increases. For P1, the maximum deflection of the curve is at around 29 °C, which coincides with the critical temperature of the polymeric network attached to a surface, as measured by force spectroscopy [34].

It is worthwhile to note that if the cationic and much more hydrophilic radical CAT1 is used as spin probe instead of TEMPO, no splitting of the high field line and hence no distinct second spin probe species can be observed. All EPR parameters (the rotational correlation time of $\tau_c = (2.3 \pm 0.5) \cdot 10^{-11}$ s and the hyperfine coupling) suggest that CAT1 is not incorporated into the collapsing regions but rather migrates completely into the aqueous phase due to its charged and hydrophilic ammonium moiety. It thus serves well for comparison purposes with respect to a chemical structure-properties relationship. Apparently, the electrostatic interaction between negatively charged MAA units and the positively charged CAT1 does not outweigh the hydrophilicity of the latter during the collapse.

To investigate the influence of polymer composition, specifically the effect of the carboxylic acid groups, hydrogel P2 without MAA units was synthesized and measured. Note that previous studies did not reveal significant differences in the collapse behavior of microscopic hydrogel films depending on whether MAA was present in the polymer or not [22]. Hence, methacrylic acid as ionic compound is not needed to prevent the skin effect. This hydrogel shows a distribution of hydrophilic and hydrophobic regions similar to P1 (shown in Fig. 5.2b). The deflection point of the collapse curve is slightly increased to 32 °C. This increase is most likely due to the absence of MAA units in the polymer backbone.

The modified chemical structure has a large impact on the hyperfine coupling constants of the two spectral components. While a_A remained almost unchanged (47.2 ± 0.2 MHz vs. 47.3 ± 0.1 MHz in P1), a_B was increased significantly from 43.2 ± 0.1 to 44.1 ± 0.2 MHz. This indicates a higher polarity of the collapsed network regions in accordance with the increased critical temperature, even though the opposite is expected due to the absence of polar acid groups in the polymer. Interestingly, this effect can be reversed by adding HCl (0.05 M) as external acid, which lowers a_B reversibly to 43.2 ± 0.1 MHz. This suggests that a more polar medium (MAA in the hydrogel or HCl in the solvent water) allows for a better phase separation. In addition, ^{1}H-NMR measurements indicate that this effect may be due to specific interactions of the amide groups with the acidic protons in the solvent or with the carboxylic acid groups in the polymer. In the collapsed polymer P1, these groups may form hydrogen-bonded pairs with a substantially reduced hydrophilicity.

5.2.2 *Chemical Decomposition of Spin Probes in Hydrophilic Regions of the Hydrogel*

Besides the specific spectral variations due to the change of the *physical* interactions of the spin probe with its environment, the overall EPR spectral signal intensity of the spin probes in the crosslinked polymer P1 significantly decreased during the

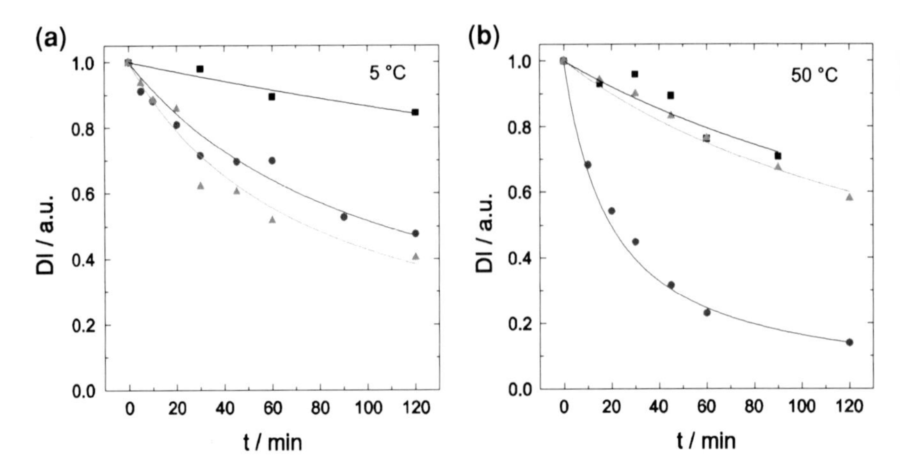

Fig. 5.3 Chemical decay of TEMPO in different environments at (**a**) 5 °C and (**b**) 50 °C, ■ 0.05 M HCl solution, ● P2 gel in 0.05 M HCl, ▲ P1 gel in pure MilliQ water. The double integral (DI) of the EPR spectrum serves as a measure for the concentration of the nitroxide. The curves are normalized to $t = 0$ to fit a second order reaction decay $\mathrm{DI} = (kt + 1)^{-1}$. Reprinted with permission from [44]. Copyright 2008 WILEY–VCH Verlag GmbH & Co. KGaA, Weinheim

series of measurements at different temperatures (typically one hour), which clearly indicates a *chemical* decomposition of the spin probes. It is well known that a disproportionation of nitroxides is catalyzed by strong mineralic acids [35]

$$2\ R_2NO^{\cdot} + H^+ \rightarrow R_2NOH + R_2N^+ = O. \quad (5.2)$$

Based on this fact, different polymer and reference samples containing spin probes were prepared and their EPR signal was recorded as a function of time at 5 and 50 °C, well below and above T_C. The concentration of the spin probe is measured by the double integrals (DI) of the CW EPR spectra. In Appendix A.3, detailed data for all investigated systems (hydrogels P1, P2, P3 and references) is provided. The most significant curves are shown in Fig. 5.3. All curves can be fitted to a second order decay $\mathrm{DI} = (kt + 1)^{-1}$ as expected for a bimolecular reaction such as the anticipated disproportionation. The extracted decay rates for all systems are summarized in Fig. 5.4.

At 5 °C, i.e. at a temperature well below T_C, *no* or *only a slight* decrease (up to 10% in 2 h) of the double integral is observed when TEMPO is dissolved in 25 mM phosphate buffer (pH 7.4), 0.05 M HCl, non-crosslinked P1 (10% in H_2O), or in a buffer solution of crosslinked hydrogel P2. In these solvents, a temperature increase to 50 °C only leads to a slight increase of the TEMPO decay (up to 20% in 2 h). The enhanced decay rates can be assigned to the increased thermal activity.

In contrast, a very strong enhancement in spin probe reactivity is observed already at 5 °C in crosslinked hydrogel P1 swollen with water and in crosslinked hydrogels P2 and P1 swollen with 0.05 M HCl solution. The concentration *decreases significantly* to about 50% of its original value within two hours.

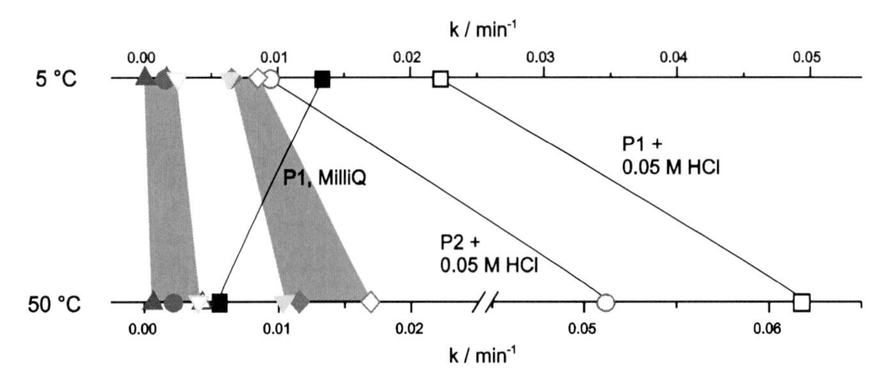

Fig. 5.4 Chemical decay rates *k* of TEMPO in the studied systems, showing the influence of the temperature with respect to the catalytic function of the hydrogels: A 25 mM phosphate buffer solution (pH 7.4) ▲, a 0.05 M HCl solution △, P2 gel swollen with 25 mM buffer solution ●, and a 10 wt% non-crosslinked P1 solution ▽ show minimal catalytic activity, which increases slightly at 50 °C. Carboxylic acid groups containing hydrogels in buffer solutions, i.e. P3 (25 mM buffer) ▼, P1 (25 mM buffer) ◆, and P1 (330 mM buffer) ◇, show an enhanced catalytic activity, which increases at 50 °C. Hydrogels swollen with 0.05 M HCl solution (P2 gel ○ and P1 gel □) experience a drastic increase in activity in the collapsed state while the opposite is true for ■ P1 gel in pure water. Reprinted with permission from [44]. Copyright 2008 WILEY-VCH Verlag GmbH & Co. KGaA, Weinheim

At 50 °C, the crosslinked hydrogel systems P1 and P2, both exposed to aqueous HCl, are responsible for an even *much faster* decay than at 5 °C. The concentration of the nitroxide is reduced to only 20% of its original value within 60 min.

Surprisingly and in contrast to the rapid decay at 5 °C, the disproportionation in the collapsed, crosslinked hydrogel P1 exposed to pure water is *much slower* at 50 °C and merely resembles that of a free aqueous HCl solution. This effect can be reversed by addition of aqueous HCl to the system. Thus, the reactivity is substantially enhanced to a similar rate as in the collapsed P2 network with HCl. This specific characteristic of the hydrogel P1 with polymer-bound carboxylic acid groups is discussed separately below in Sect. 5.3.3.

Two main factors trigger the catalytic activity: (1) the availability of acidic protons and (2) a crosslinked polymer network, as discussed in detail in Sects. 5.3.2 and 5.3.3. None of these individual factors *alone* has a strong influence on the reactivity for the spin probe decay. A maximal catalytic acitivity is only achieved if both factors are combined.

At 5 °C, the combined properties of hydrogel P1 trigger its catalytic activity. On the one hand, the hydrogel network provides a spatially confined environment and specific reaction sites where two TEMPO molecules may reside in close proximity. On the other hand, carboxylic acid units catalyze the disproportionation reaction. These acid units need not necessarily be incorporated into the polymer chain directly, as shown by the addition of an external acid like HCl to the intrinsically acid-free polymer network P2 (or to the collapsed P1 at 50 °C), yielding a qualitatively comparable reaction enhancement.

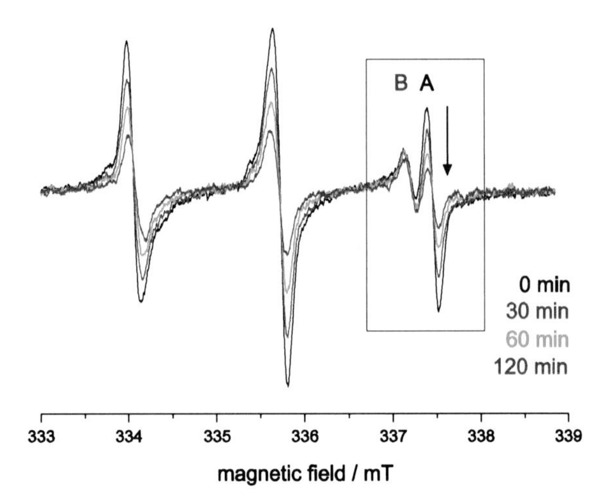

Fig. 5.5 CW EPR spectra of TEMPO placed in a 25 mM phosphate buffer solution of the collapsed gel P1 at 50 °C and different reaction times: 0 min (*black*), 30 min (*blue*), 60 min (*green*), 100 min (*red*). The signal of the hydrophilic species A decreases drastically while the absolute spectral contribution of species B remains unchanged. Reprinted with permission from [44]. Copyright 2008 WILEY–VCH Verlag GmbH & Co. KGaA, Weinheim

At 50 °C, additional heterogeneities and interfaces are introduced in the collapsed gel. A very intriguing detail of the spin probe reaction at 50 °C is displayed in Fig. 5.5. While species A, i.e. TEMPO in a hydrophilic environment, degrades rapidly, the signal of species B (TEMPO in a hydrophobic environment) remains constant over two hours. This selectivity is observed for all hydrogel samples under investigation. Besides an intriguing implication for the catalytic activity, which is discussed in detail in Sect. 5.3.2, a second conclusion can be drawn from this observation.

The selective decay of only one species implies that the spin probes do not exchange over a time scale of at least two hours. Thus, the inhomogeneities, in which the spin probes reside, are *static* and do not fluctuate in a remarkably long time frame. A thermoresponsive gel contains both static and dynamic inhomogeneities [15]. Static inhomogeneities originate from the crosslinking process, while dynamic fluctuations are observed near the critical temperature at the onset of gelation [36–38]. In this study, the static inhomogeneities are selectively sampled by the spin probe while dynamic fluctuations remain invisible.

5.3 Discussion

5.3.1 The Hydrogel Collapse on a Molecular Level

Macroscopic characterization methods such as volumetric determination [11], turbidity measurements [39], force spectroscopy [34], and optical waveguide spectroscopy (OWS) [22, 38] suggest a temperature range ΔT of the collapse

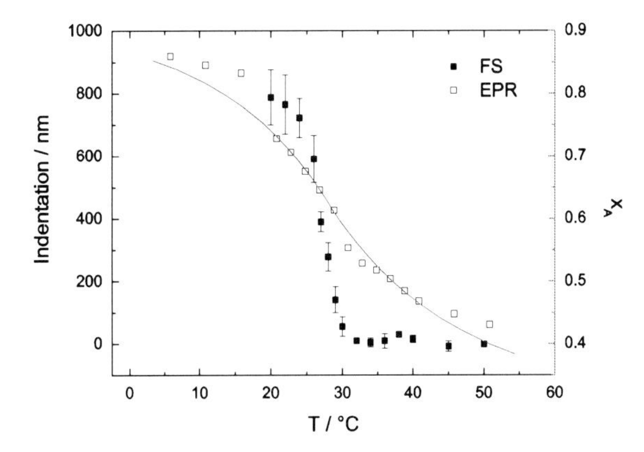

Fig. 5.6 Comparison of different phase transition measurements on hydrogel P1. A collapse curve, measured by force spectroscopy [34], is presented, which relates to changes of macroscopic properties (*black closed squares*). It is compared to the EPR collapse curve (*blue open squares*), which samples changes on the molecular scale on hydrogel. The macroscopic curve shows a narrow transition width of $\Delta T \approx 10$ K while a much broader temperature range of $\Delta T \geq 40$ K is observed on the molecular level

transition of about 10 K. In contrast, a substantially broadened transition range of $\Delta T_{EPR} \geq 40$ K is observed by the spin probe technique presented here (Fig. 5.6). This apparent discrepancy is due to the sensitivity of spin probes for changes of the environment on the molecular level. This allows the detection of temperature-dependent changes on the spatial structure in the polymer chains even while the macroscopic features are not significantly altered.

In an inhomogeneous hydrogel—where dense and open regions coexist—the collapse of dense regions is thermodynamically favored over the collapse of the open regions at low temperatures. Since the corresponding volume change is small, the open regions remain and may partially compensate the collapse-induced strain in the dense regions. Thus, the local collapse cannot be detected macroscopically but has significant impact at the molecular scale. The same is true for high temperatures, as the remaining swollen regions are very small due to the strain exerted on them by the collapse of the surrounding matrix. This interpretation is supported by a recent force spectroscopy study on the same hydrogel system [34]. Independent of macroscopic changes, hydrophobic interactions were found to increase with temperature, which is indicative of inhomogeneity formation far beyond macroscopic limits. Also, nonradiative energy transfer studies on pNiPAAm solutions revealed a gradual shrinking of the polymer coil into a collapsed state below the LCST, which was then followed by aggregation [40, 41].

From theoretical considerations [36], there are two possible ways to get from an expanded to a collapsed hydrogel, as depicted in Fig. 5.7. In the upper path, all network meshes gradually decrease in size in an isotropic manner until the fully collapsed structure is obtained. This spin probe study strongly indicates that this picture does not correctly describe the collapse on a molecular scale for the given

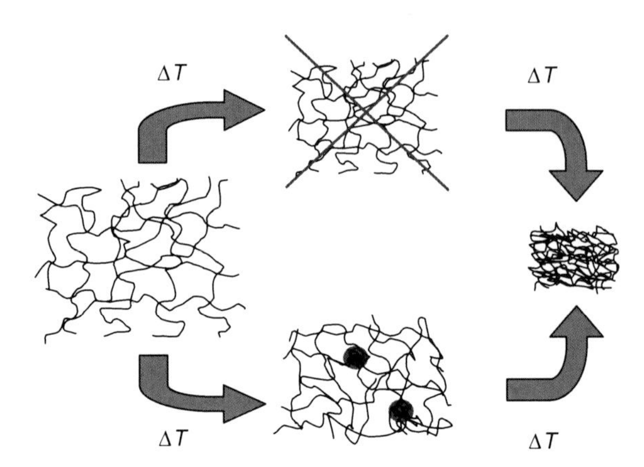

Fig. 5.7 Schematic picture of the hydrogel structure during the collapse: The upper path assumes isotropic shrinking of the network meshes, which is not supported by the experimental data. Instead, a microphase separation takes place with coexisting swollen and collapsed regions during the phase transition. Reprinted with permission from [44]. Copyright 2008 WILEY-VCH Verlag GmbH & Co. KGaA, Weinheim

hydrogel systems. Rather, individual entities seem to collapse completely without having a major influence on those regions that are still in the expanded state (Fig. 5.7, lower part). Hence, a microphase separation is observed on the molecular level. This observation is supported by MD simulations by Rabin and coworkers on two-dimensional model hydrogels. In this study, the authors observed locally collapsed structures during the collapse process [42].

In essence, by changes of the *physical* interactions of the spin probe with the hydrogel during the temperature-induced collapse, structural changes of the gel are monitored in the EPR spectrum showing an inhomogeneous collapse process with coexisting hydrophilic and hydrophobic regions.

5.3.2 Hydrogel Inhomogeneities on the Nanoscale Lead to Nanoreactors and Nanoshelters

The detailed series of experiments concerning the chemical decomposition of spin probes (Sect. 5.2.1) reveals a remarkable feature of thermoresponsive hydrogels.

Since at 5 °C, well below T_C, nearly all spin probes are located in a hydrophilic environment within the hydrogel, the measured chemical decay curves suggest that such regions can be described as *nanoreactors*, in which the acid-catalyzed bimolecular spin probe decay takes place at a strongly enhanced reaction rate.

The time dependence of the CW EPR spectral intensity shown in Fig. 5.5 for temperatures above T_C can be interpreted such that collapsed hydrophobic regions act as *nanoshelters*, protecting the TEMPO molecules therein from access by reagents (other TEMPO molecules) and/or catalysts (active protons and hydrogen-bonded structures).

Thus, in the collapsed hydrogels one observes the simultaneous occurrence of nanoshelters prohibiting the spin probe decay and nanoreactors, hydrophilic hydrogel regions, in which the reaction is significantly accelerated. The coexistence of such nano-inhomogeneities not only in structure but also in function bears striking resemblance with functional proteins and enzymes, in which catalytic activity crucially depends on a sophisticated interplay of hydrophilic and hydrophobic substructures, in this context called *active sites* and *hydrophobic clefts*. The chosen terminology (nanoshelters and nanoreactors) is derived from the knowledge that EPR spectroscopy is sensitive towards spatial rearrangements on the nanometer scale combined with the fact that a mesh size of several nanometers was observed in recent studies with photon- and fluorescence correlation spectroscopy [27, 28].

In summary, the observed time dependent drop of the overall EPR signal intensity is a clear indication of a *chemical* spin probe decomposition, which is catalyzed by the hydrogel network and depends sensitively on both structure and composition of the system.

5.3.3 Nanoreactors With Localized Acid Groups

As mentioned before, one striking anomaly to the above described behavior of reactivity enhancement with increasing temperature is apparent in Fig. 5.3 and in the schematic presented in Fig. 5.4. At 50 °C, the overall reaction rate of TEMPO in water-swollen hydrogel P1 with carboxylic acid substituents is substantially *slower* than at 5 °C. In order to understand this anomalous feature, a decay curve was recorded at an intermediate temperature of 35 °C, only slightly above the critical temperature (data not shown). This decay curve displays a TEMPO decay reaction rate that is intermediate between that at 5 and 50 °C, but closer to the 50 °C behavior. Hydrogel P1 contains carboxylic acid moieties and simultaneously primary amide groups with amide protons, which can form hydrogen-bonded networks. Based on this feature, a more detailed molecular picture of the nano-inhomogeneities of the hydrogel is derived. Its implications for the spin probe decay reaction are schematically shown in Fig. 5.8.

When free protons are introduced into the systems by addition of 0.05 M HCl, the decay reaction is catalyzed by two factors: the spatial confinement of more than one spin probe in remaining hydrophilic regions of the hydrogel network, and the availability of highly dissociated (i.e. mobile) protons. The same is true for

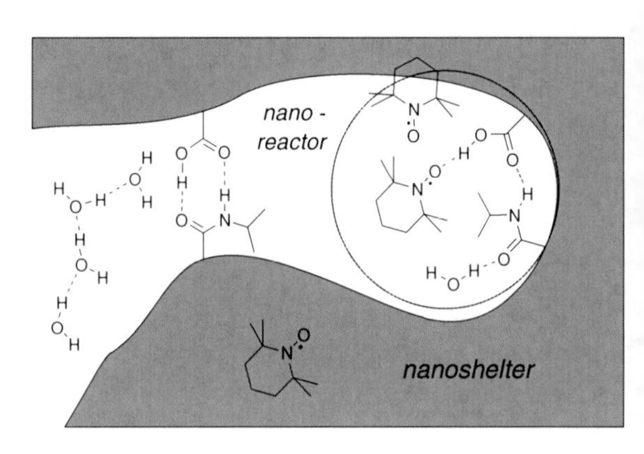

Fig. 5.8 Schematic illustration of the nano-inhomogeneities in the studied hydrogels. The dotted circle marks the part of the nanoreactor that functions in analogy to an enzymatic active site. Reprinted with permission from [44]. Copyright 2008 WILEY-VCH Verlag GmbH & Co. KGaA, Weinheim

hydrogel P1 in a buffer solution, since the protons of the carboxylic groups are mainly transferred to mobile buffer molecules. Due to the amphiphilicity of the TEMPO spin probes with the polar nitroxide group and the apolar hydrocarbon ring, this confinement probably takes place at the interface between nano-inhomogeneities in the hydrogel network. In analogy to most catalytic agents in heterogeneous catalysis, the large total area of this interface may be a key factor with respect to the reaction acceleration.

When hydrogel P1 is used without addition of HCl, one is faced with a system that nominally contains fewer acidic groups that are only dissociate to a degree much lower than unity ($pK_A \sim 5$). Yet, at 5 °C the spin probe decay is catalyzed even more efficiently than in a hydrogel with added HCl, while the catalytic effect becomes weaker with increasing temperature. As shown above, the catalytic activity depends on the availability of acidic protons, thus a possible explanation for the enhanced reactivity at low temperatures may lie in the extended hydrogen bonded structure between the carboxylic acid and amide groups of hydrogel P1, which activates the carboxylic acid protons. The mutual influence between the carboxylic acid and amide groups is corroborated by the significant chemical shift of the amide proton in ^{1}H-NMR compared to polymers without carboxylic acid groups.

In other words, most of the carboxylic acid protons in gel P1 are localized, i.e. not fully dissociated, but activated by hydrogen bonding with the amide groups. Due to the confined spatial structure at the hydrophilic interfaces of the hydrogel, they are in proximity to the spin probes. Hence, the spin probe decay in these systems does not directly depend on the number of mobile protons, i.e. the bulk pH, but rather on the existence of few but activated carboxylic acid protons at the reaction sites. Apparently, during the temperature-induced collapse this hydrogen-bonded structure is weakened, which leads to a significant decrease in catalytic activity.

Again, this feature bears analogy with active sites in functional biological macromolecules. As a system with a remarkably simple chemical composition, the crosslinked hydrogel P1 shows a highly complex spatial structure and chemical functionality.

5.4 Conclusions

The above described experiments show that an amphiphilic paramagnetic molecule serves as a probe to sample the structural changes of a temperature-induced hydrogel collapse. This is documented by the coexistence of two dynamic species, which change their relative ratios with the temperature and consequently with the swelling state of the responsive hydrogel. The EPR measurements reveal that the hydrogel does not shrink isotropically during the thermally induced collapse. Rather, a microphase separation takes place over a broad temperature range of at least 40 K with coexisting swollen and collapsed network regions, even at temperatures where the gel does not show any macroscopic volume changes.

Small spin probe molecule may represent a substitute for drugs and other active species that are incorporated in and diffuse through the hydrogel network. The obtained results might have substantial implications for the application and design of responsive hydrogels in sensors or drug-release systems. Specifically, the inhomogeneous structural change during the phase separation and the entrapment of small amphiphilic molecules in the hydrophobic regions will have to be considered in such applications.

A second remarkable feature is the chemical decomposition of the spin probe in the hydrogel network. By varying the chemical composition of the polymer and the liquid medium (pH and buffer), the temperature-dependent spin probe decay was monitored and a structural model of the specific requirements for the catalytic activity of the hydrogel network was deduced. In summary, an inhomogeneous polymer network with hydrophilic cavities is required, which contain active protons and can accommodate more than one spin probe. In this respect, hydrogel P1 with carboxylic acid functions and primary amide groups represents—on a very basic level—a temperature-responsive enzyme analogue that catalyzes the spin probe decay at temperatures below the phase transition. Above T_C, these hydrophilic regions coexist with hydrophobic cavities that protect the spin probe from any decay reaction. Such a complex catalytic behavior is remarkable when taking into consideration that these hydrogels consist of relatively simple, statistical binary or ternary copolymers with one comonomer exhibiting an LCST. Yet, as a consequence of the combination of spatial confinement and the existence of protons, either in activated H-bonded networks or as fully mobile bulk protons, they achieve chemical reaction catalysis comparable to functional biomacromolecules that are optimized for a specific type of reaction.

5.5 Materials and Methods

Materials. *N*-Isopropylacrylamide (NiPAAm, 99%, Aldrich) was recrystallized from toluene/hexane (ratio 1:4). Methacrylic acid (MAA, 99%, Aldrich) was distilled prior to use. 4-Methacryloyloxy-benzophenone (MABP) was prepared according to the literature [25]. Dioxane was distilled over CaH_2 and dried over a

Table 5.1 Properties of the synthesized polymers

	Monomer composition						
	NiPAAm/DEAAm	MAA	MABP	M_w/g mol^{-1} [a]	M_w/M_n [a]	T_g/°C [b]	Yield/%
P1	94 (NiPAAm)	5	1	206,000	2.0	142	83
P2	99 (NiPAAm)	–	1	180,000	2.5	139	84
P3	94 (DEAAm)	5	1	81,600	3.0	104	88

[a] determined by GPC at 60 °C in DMF, standard: PMMA
[b] measured by DSC, heating rate: 10 K min^{-1}

molecular sieve (4 Å). 2,2'-Azobis(isobutyronitrile) (AIBN, 98%, Acros) was recrystallized from methanol. *N,N*-Diethylacrylamide (DEAAm, 99%, Polysciences Inc.), 2,2,6,6-tetramethylpiperidine-1-oxyl (TEMPO, 98%, Aldrich) and the corresponding 4-trimethylammonium chloride derivative (CAT1, Molecular Probes Inc.) were used as received. Distilled water was further purified by a MilliQ System (Millipore) to achieve a resistivity of 18.2 MΩ cm.

p(NiPAAm-co-MAA-co-MABP) (P1) and p(NiPAAm-co-MABP) (P2). NiPAAm (3 g, 26.5 mmol, 94 eq (P1), 99 eq (P2)), MAA (114 mg, 1.32 mmol, 5 eq, P1) and MABP (71 mg, 0.27 mmol, 1 eq, P1 + P2) were polymerized with AIBN (20 mg, 0.12 mmol) as initiator under exclusion of air and moisture. The reaction took place in 20 mL dioxane at 60 °C for 24 h [22]. The polymers were precipitated in 200 mL diethyl ether and purified by re-precipitation from ethanol into diethyl ether. They were freeze-dried from *tert*-butanol in vacuo. The yield was around 83%. ^{1}H-NMR (250 MHz, d^4-MeOH): δ/ppm = 0.9–1.2 (m, -$\mathbf{CH_3}$ NiPAAm, MAA, MABP), 1.3–1.9 (m, $\mathbf{CH_2}$ backbone), 1.7–2.3 (m, **CH** backbone, NiPAAm), 3.7–4.1 (m, **CH**$(CH_3)_2$, NiPAAm), 7.2–8.1 (m, C-$\mathbf{H_{arom}}$ + C(= O)**NH**···**COOH** in P1). The broad peak in the aromatic region is additionally enlarged by NiPAAm amide protons, which are hydrogen bonded to the carboxylic acid groups of MAA. Via integration of selected NMR peaks, the polymer composition was found to resemble the composition of the monomers.

p(DEAAm-co-MAA-co-MABP) (P3). The general synthetic procedure of P1 was followed, replacing NiPAAm with DEAAm (3 g, 23.6 mmol). The solvents for precipitation had to be adjusted to a mixture of diethyl ether and hexane (1/1). ^{1}H-NMR (250 MHz, d^4-MeOH): δ/ppm = 0.8–1.5 (m, -$\mathbf{CH_3}$ DEAAm, MAA, MABP), 1.5–1.2 (m, $\mathbf{CH_2}$ backbone), 2.3–2.9 (m, **CH** backbone, DEAAm), 2.9–3.8 (m, $\mathbf{CH_2}CH_3$, DEAAm), 6.8–8.0 (m, C-$\mathbf{H_{arom}}$).

The properties of the obtained polymers are listed in Table 5.1.

Sample Preparation. To achieve homogeneously crosslinked hydrogel films, 15 wt% solutions of the polymers in ethanol were drop-cast on a slide and dried overnight in vacuo at 50 °C. Films in the range of 10–15 μm were obtained as checked with a profilometer (Tencor P-10 Surface Profiler, KLA Tencor). The polymer was crosslinked by UV irradiation (13.64 J/cm^2) with a Stratagene UV Stratalinker with a peak wavelength of 365 nm and an irradiance of 1.74 W/cm^2. The crosslinked films were abraded with a scalpel, filled into the EPR sample canules (i.d. 0.8 mm) and swollen with a 0.2 mM solution of the spin probes,

TEMPO or CAT1, in MilliQ water. Excess water not bound to the hydrogel was removed with a paper tissue to eliminate signals from freely rotating spin probes in the bulk aqueous phase.

CW EPR Measurements. The spectra were recorded with a Miniscope MS200 (Magnettech, Berlin, Germany) benchtop spectrometer working at X-band (~9.4 GHz). Typical experimental parameters were a modulation amplitude of 0.04 mT and a sweep width of 6 mT. The microwave frequency was recorded with a frequency counter, model 2101 (Racal-Dana). The temperature was adjusted with the temperature control unit TC H02 (Magnettech), providing an electronic adjustment in steps of 0.1 °C in the range of −170 to 250 °C. The sample temperature was checked with a Pt100 and small deviations to the displayed temperature were corrected.

Data Analysis and Interpretation. The spectra were simulated with a Matlab program, which utilizes the Easyspin software package for EPR (see Sect. 2.7) [43]. Simulations were performed using either the simulation routine for fast or intermediate/slow motion. Both routines provided virtually identical hyperfine coupling constants and x_A-values (cf. Eq. 5.1). The rotational correlation times are extracted from intermediate motion simulations. The evolution of the spin probe concentration over time was determined by double integration of the spectra. The most reliable method to account for baseline effects while achieving comparability was a zeroth order baseline correction.

References

1. Qiu Y, Park K (2001) Adv Drug Delivery Rev 53:321–339
2. Smith AE, Xu XW, McCormick CL (2010) Prog Polym Sci 35:45–93
3. Hoffman AS (2002) Adv Drug Delivery Rev 54:3–12
4. Hoffman AS, Stayton PS, Bulmus V, Chen GH, Chen JP, Cheung C, Chilkoti A, Ding ZL, Dong LC, Fong R, Lackey CA, Long CJ, Miura M, Morris JE, Murthy N, Nabeshima Y, Park TG, Press OW, Shimoboji T, Shoemaker S, Yang HJ, Monji N, Nowinski RC, Cole CA, Priest JH, Harris JM, Nakamae K, Nishino T, Miyata T (2000) J Biomed Mater Res 52:577–586
5. Peppas NA, Hilt JZ, Khademhosseini A, Langer R (2006) Adv Mater 18:1345–1360
6. Richter A, Kuckling D, Howitz S, Gehring T, Arndt KF (2003) J Microelectromech S 12:748–753
7. Chung JE, Yokoyama M, Yamato M, Aoyagi T, Sakurai Y, Okano T (1999) J Controlled Release 62:115–127
8. Soppimath KS, Liu LH, Seow WY, Liu SQ, Powell R, Chan P, Yang YY (2007) Adv Funct Mater 17:355–362
9. Schild HG (1992) Prog Polym Sci 17:163–249
10. Hirotsu S, Hirokawa Y, Tanaka T (1987) J Chem Phys 87:1392–1395
11. Yu H, Grainger DW (1993) J Appl Polym Sci 49:1553–1563
12. Lin SY, Chen KS, Liang RC (1999) Polymer 40:2619–2624
13. Percot A, Zhu XX, Lafleur M (2000) J Polym Sci Part B: Polym Phys 38:907–915
14. Harmon ME, Kuckling D, Pareek P, Frank CW (2003) Langmuir 19:10947–10956
15. Ikkai F, Shibayama M (2005) J Polym Sci Part B: Polym Phys 43:617–628
16. Kariyo S, Küppers M, Badiger MV, Prabhakar A, Jagadeesh B, Stapf S, Blümich B (2005) Magn Reson Imag 23:249–253

17. Hinderberger D, Jeschke G (2006) In: Webb GA (ed) Site-specific characterization of structure and dynamics of complex materials by EPR spin probes in modern magnetic resonance, Springer, Berlin
18. Hinderberger D, Schmelz O, Rehahn M, Jeschke G (2004) Angew Chem Int Ed 43:4616–4621
19. Harvey RD, Schlick S (1989) Polymer 30:11–16
20. Rex GC, Schlick S (1987) Polymer 28:2134–2138
21. Winnik FM, Ottaviani MF, Boßmann SH, Pan W, Garcia-Garibay M, Turro NJ (1993) Macromolecules 26:4577–4585
22. Beines PW, Klosterkamp I, Menges B, Jonas U, Knoll W (2007) Langmuir 23:2231–2238
23. Kuckling D, Harmon ME, Frank CW (2002) Macromolecules 35:6377–6383
24. Xue W, Hamley IW, Huglin MB (2002) Polymer 43:5181–5186
25. Toomey R, Freidank D, Rühe J (2004) Macromolecules 37:882–887
26. Aguilar MR, Elvira C, Gallardo A, Vázquez B, Román JS (2007) In: Ashammakhi N, Reis R, Chiellini E (eds) Smart polymers and their applications as biomaterials in topics in tissue engineering, vol 3
27. Gianneli M, Beines PW, Roskamp RF, Koynov K, Fytas G, Knoll W (2007) J Phys Chem C 111:13205–13211
28. Gianneli M, Roskamp RF, Jonas U, Loppinet B, Fytas G, Knoll W (2008) Soft Matter 4: 1443–1447
29. Atherton NM (1993) Principles of electron spin resonance. Ellis Horwood, New York
30. Knauer BR, Napier JJ (1976) J Am Chem Soc 98:4395–4400
31. Ondar MA, Grinberg OY, Dubinskii AA, Lebedev YS (1985) Sov J Chem Phys 3:781–792
32. Note that these measured hyperfine coupling constants are slightly altered by inaccuracies in the magnetic field steps. In the present case, the apparent a_{iso} values are decreased by roughly 1 MHz with respect the real values, as checked by comparison with reported values of nitroxides in water (cf. Ref. [30]). The corrected value a_B was compared to literature data of nitroxides in various solvents
33. Mukerjee P, Ramachandran C, Pyter RA (1982) J Phys Chem 86:3189–3197
34. Junk MJN, Berger R, Jonas U (2010) Langmuir 26:7262–7269
35. Golubev VA, Kozlov YN, Petrov AN, Purmal AP (1992) In: Zhdanov RI (ed) Catalysis of redox processes by nitroxyl radicals in bioactive spin labels, Springer, Berlin
36. Shibayama M (1998) Macromol Chem Phys 199:1–30
37. Shibayama M, Takata S, Norisuye T (1998) Phys A 249:245–252
38. Junk MJN, Anac I, Menges B, Jonas U (2010) Langmuir 26:12253–12259
39. Li Y, Wang GN, Hu ZB (1995) Macromolecules 28:4194–4197
40. Winnik FM (1990) Polymer 31:2125–2134
41. Schild HG, Tirrell DA (1992) Macromolecules 25:4553–4558
42. Peleg O, Kröger M, Hecht I, Rabin Y (2007) Europhys Lett 77:58007
43. Stoll S, Schweiger A (2006) J Magn Reson 178:42–55
44. Junk MJN, Jonas U, Hinderberger D (2008) Small 4:1485–1493

Chapter 6
Thermoresponsive Spin-Labeled Hydrogels as Separable DNP Polarizing Agents

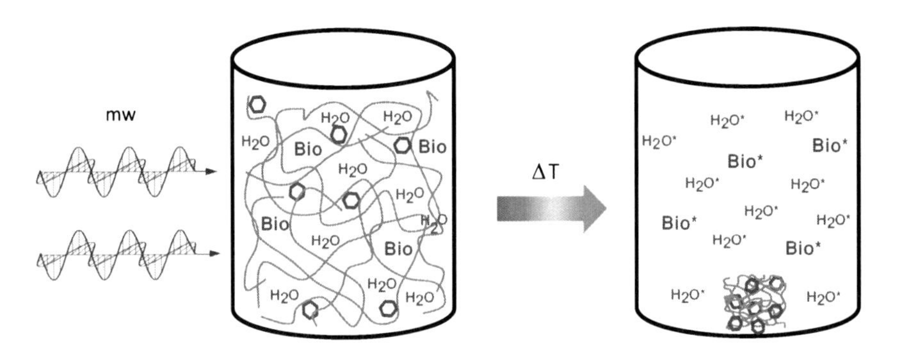

Dynamic nuclear polarization (DNP) is a commonly applied NMR hyperpolarization technique, which is based on the polarization transfer from electron spins to nuclear spins. While DNP allows a significant enhancement of NMR signals by several orders of magnitude, major drawbacks of the method include enhanced nuclear relaxation times due to the presence of unpaired electrons and the toxicity of radicals, which is the limiting factor for in vivo applications in magnetic resonance imaging. Thus, an efficient separation of the polarization agent is a key requirement for possible applications in medicine. In this chapter, the application of spin-labeled thermoresponsive hydrogels as polarizing agents for dynamic nuclear polarization is explored. In the approach presented here, the thermally triggered macroscopic phase separation is utilized to efficiently separate the polarizing agents that are embedded in the hydrogel matrix from the molecules of interest. A radical-free solution is obtained, which exhibits a prolonged lifetime of the hyperpolarization.

M. J. N. Junk, *Assessing the Functional Structure of Molecular Transporters by EPR Spectroscopy*, Springer Theses, DOI: 10.1007/978-3-642-25135-1_6,

6.1 Introduction

Nuclear magnetic resonance (NMR) is the most widely used source for structural and dynamic information of molecules and molecular assemblies. Because of its versatility, NMR is applied in a large spectrum of scientific disciplines ranging from materials science to biomedical research. However, the method suffers from a major drawback. The small magnetogyric ratios of the nuclear spins only lead to a small energetic difference of the spin states in an external magnetic field. Thus, a low thermal polarization is obtained, which leads to a low sensitivity (cf. Sect. 2.2.3). For ^1H, the nucleus with the highest magnetogyric ratio, a thermal polarization as small as $9.0 \cdot 10^{-6}$ is obtained at a magnetic field of 16.4 T (700 MHz) and at 293 K (cf. Eq. 2.6). The lack of sensitivity is even more prominent for biologically relevant heteronuclei like ^{13}C and ^{15}N, which exhibit far smaller magnetogyric ratios.

A higher Boltzmann population can be achieved by an increase of the magnetic field strength or a decrease of the temperature, but the variation of both parameters is limited. Beside these conventional methods to increase the signal-to-noise ratio (SNR), a variety of techniques exist, which couple the nuclear spin system to other transitions and thus achieve substantially higher polarizations than in thermal equilibrium.

For instance, the transfer of electron spin polarization of photoexcited radical pairs or triplet states causes (photo-)chemically induced dynamic nuclear polarization (CIDNP) [1–4]. Similarly, the pairwise transfer of dihydrogen in its *para*-state can be utilized to obtain parahydrogen induced polarization (PHIP) [5–8]. In another approach, the angular momentum of photons in polarized laser light is transferred to helium or xenon to produce hyperpolarized gases (optically pumped NMR) [9]. However, all methods introduced so far are restricted to a narrow range of molecules and applications.

In contrast, dynamic nuclear polarization (DNP) is a versatile hyperpolarization method. It only requires the presence of paramagnetic species in the sample since it is based on the transfer of polarization from unpaired electron spins to nuclear spins. Predicted already in 1953 by Overhauser [10] and verified only shortly afterwards by Carver and Slichter [11, 12], it has only recently found widespread application in physics, chemistry, biology, and medicine [13–19].

DNP techniques can be divided into two subgroups, in situ and ex situ DNP. In *in situ* DNP, polarization and NMR/MRI detection are implemented in the same magnet. It was applied to Overhauser DNP at high and low magnetic field strengths [20–24] and to solid-state DNP [25, 26]. In contrast, polarization and NMR/MRI detection take place in different magnets in ex situ DNP. Prominent examples for ex situ methods are shuttle DNP and dissolution DNP [27–29].

Each of these methods has its advantages and is suitable for certain applications. However, DNP is severely hampered by a key problem, which originates from the introduction of radicals to the target molecules. The presence of these radicals significantly decreases the T_1 and T_2 relaxation times of the nuclear spins.

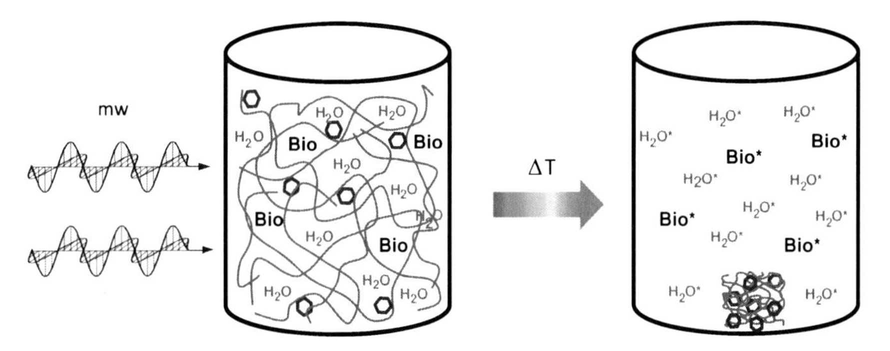

Scheme 6.1 Schematic presentation of DNP in a thermoresponsive spin-labeled hydrogel and subsequent thermally induced collapse, which results in the separation of radicals (*red hexagons*) from hyperpolarized (*) water and target biomolecules (*Bio*) [54]—Reproduced by Permission of the PCCP Owner Societies

This leads to a line broadening of the NMR signal, and, more severely, limits the time frame during which the accomplished hyperpolarization can be used. This is a major problem for the ex situ DNP methods due to T_1 relaxation during the transfer time. An additional problem is related to the toxicity of the radical. Thus, a hyperpolarized solution cannot be used for in vivo applications (at least in humans) without the radical being removed.

Hence, a fast and reliable separation of radicals and polarized material remains an important issue for an improved applicability of DNP. Two different procedures for radical separation were proposed in the literature: immobilization of the radicals in silica or gel beads combined with a continuous flow during the polarization step to separate the solute from the radicals [30, 31], and the filtration of the radicals using ion-exchange columns [28, 32]. Especially the application of gel beads manufactured from hydrophilic polymer networks seems to be promising. McCarney et al. [31] reported enhancement factors for water incorporated in a spin-labeled sepharose gel exceeding those obtained with silica-embedded radicals, which was attributed to the higher mobility of the spin labels in the water-swollen gels compared to the solid silica.

In this chapter, thermoresponsive, spin-labeled hydrophilic polymer networks are introduced as separable DNP polarizing agents. The thermally triggered macroscopic collapse above the critical temperature T_C is utilized for a fast and simple radical–solute separation after the hyperpolarization (Scheme 6.1). Due to this phase separation, incorporated water and (biological) target molecules are efficiently expelled from the radical-bearing polymer network [33, 34].

After a short introduction to Overhauser DNP, the choice of the synthesized gel systems is explained. Via CW EPR, it is proven that the radicals are covalently attached and trapped in the network. In the following, the achieved DNP enhancement factors and their dependence on the applied microwave power and temperature are discussed. The results of these first model systems are then used to derive a synthetic strategy to responsive gels with improved DNP polarization efficiencies.

6.2 Theory of Overhauser DNP

In solution, the Overhauser effect is mainly responsible for the polarization transfer between electron spins and nuclear spins. The enhancement of the NMR signal E is given by [35]

$$E = 1 - \xi f s \frac{|\gamma_S|}{\gamma_I}. \tag{6.1}$$

The maximum achievable enhancement is limited by the ratio of the magnetogyric ratios of the electron and the nucleus, i.e. theoretically, a factor of ~660 can be accomplished for protons. ξ, f ,and s are constants, which characterize the efficiency of the polarization transfer.

The coupling factor ξ is the ratio of the electron–nuclear spin cross relaxation rate and the nuclear spin relaxation rate due to the electrons and expresses the efficiency of the coupling between the electron and nuclear spin. The coupling factor ranges from +0.5 to −1, dependent on whether the coupling is due to dipolar or scalar interactions. In case of a dipolar coupled system, the relaxation mainly proceeds via the double quantum transition while the zero quantum relaxation is negligible. The opposite is true for scalar coupling. It has been shown that the dipolar coupling is the by far dominant contribution for nitroxides in solution [36, 37].

The polarization transfer in a four-level system (cf. Sect. 2.8.5 and Fig. 2.15) of dipolar coupled spins is graphically illustrated in Scheme 6.2. By saturation of the EPR transitions, the populations of the high and low energy levels are equalized. Since the double quantum relaxation pathway dominates over zero quantum relaxation, a large number of spins is transferred to the nuclear $|\beta\rangle$ manifold, which leads to NMR emission lines, i.e. to a negative signal enhancement.

The leakage factor f characterizes the ratio of the nuclear relaxation rate with and without the presence of electrons. Hence, it can be determined by the respective longitudinal relaxation times of the sample,

$$f = 1 - \frac{T_1}{T_{1,0}}. \tag{6.2}$$

The saturation factor s is a measure for the quality of the saturation of the EPR lines, which is achieved by microwave irradiation. It is given by

$$s = \frac{S_0 - \langle S_z \rangle}{S_0}. \tag{6.3}$$

$\langle S_z \rangle$ is the residual electron magnetization under saturation conditions, S_0 the initial magnetization at thermal equilibrium. If the three different EPR transitions of a nitroxide are regarded as independent, only one line of the spectrum can be saturated and the maximum achievable Overhauser enhancement is limited to −110 for ^{14}N-nitroxides. However, an additional partial saturation of the other

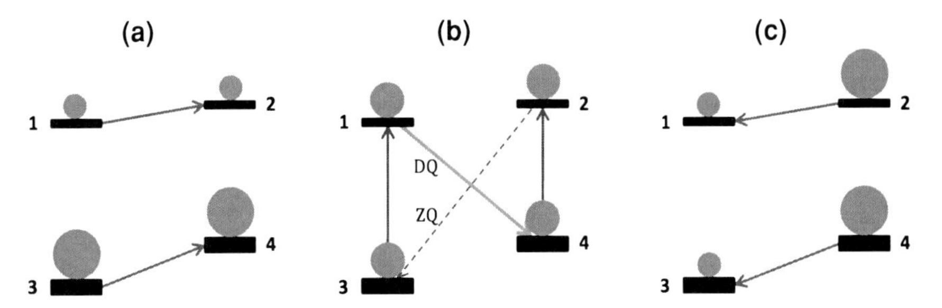

Scheme 6.2 Schematic illustration of the Overhauser effect in the framework of the four-level scheme (cf. Fig. 2.15). **a** Boltzmann population in the thermal equilibrium. The NMR transitions are marked by red lines. **b** By saturation of both EPR transitions (*blue*) and a dominant relaxation pathway via the double quantum transition (*orange*), a hyperpolarized state **c** is achieved which leads to a high population of the level with the nuclear spin manifold $|\beta\rangle$ and to a negative enhancement of the NMR signal

EPR lines can be achieved due to spin relaxation. In the limit of fast rotation, Heisenberg spin exchange is mainly responsible for an electron exchange between the $m_{\rm I}$ manifolds [38]. At low rotational mobility, the nitrogen nuclear relaxation rate exceeds the relaxation rate of the electron spin providing an averaged view on different $m_{\rm I}$ manifolds [39, 40]. Thus, enhancements exceeding the value of -110 can be achieved (see below).

6.3 Results and Discussion

6.3.1 Strategy for the Preparation of Thermoresponsive Spin-Labeled Hydrogels

Responsive hydrogels are most commonly synthesized via a free radical polymerization in water with potassium persulfate as initiator. With this fast and simple method, macroscopic water-swollen gels are obtained. However, it is difficult to completely remove the incorporated water, if a water-free environment is required for a chemical modification of the gel. Further, the substantially decreased diffusion of the reactants in the gel will result in a slow reaction rate.

Based on the experience with photocrosslinked systems, P1 (introduced in Chap. 5) was taken as starting point. Based on *N*-isopropylacrylamide (NiPAAm), it is prepared under non-aqueous conditions and can be photocrosslinked as thin film, which is easily accessible by reactants [41]. Its methacrylic acid (MAA) units do not only prevent the skin effect, but are reactive groups, which can be utilized to chemically modify the gel. This strategy was previously applied to covalently attach amino-functionalized dyes to the polymer [42]. Adapting the reported method, the carboxylic acid groups were first converted into active esters, which were then reacted with an amino-functionalized nitroxide.

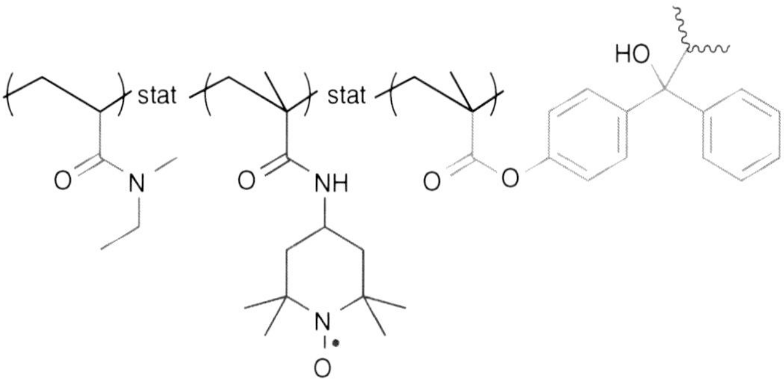

P5	94	:	5	:	1
P15	84	:	15	:	1

Scheme 6.3 Molecular structure and composition of the spin-labeled thermoresponsive hydrogel systems based on *N*,*N*-ethylmethylacrylamide (EMAAm). The numbers denote the ratios of the original monomers in the polymerization feed

Since a high radical concentration is needed for DNP, the content of MAA in the initial polymerization feed was increased to 15%. Thus, the overall radical concentration in the swollen state could be raised to ~6.8 mM (compared to ~1 mM for 5% MAA). However, first DNP experiments only resulted in a maximum ^{1}H NMR signal enhancement of $E = -16.2$, which was far less than expected. Soon it became clear, that the low enhancement was related to a partial collapse of the hydrogel due to a microwave-induced heating near or above its critical temperature of 32 °C (see Sects. 6.3.3 and 6.3.4 for a detailed discussion) [34, 43]. Thus, *N*,*N*-ethylmethylacrylamide (EMAAm) was chosen as thermoresponsive unit, which exhibits an elevated lower critical solution temperature of 63 °C and is less affected by the microwave irradiation [44, 45]. The molecular composition of the hydrogels discussed in this chapter is shown in Scheme 6.3.

6.3.2 CW EPR Characterization of the Spin-Labeled Hydrogels

First, CW EPR was used to check if the spin labels are covalently attached to the hydrogels. In Fig. 6.1, the CW EPR spectra of P5 and P15 are shown at 5 °C. Broad EPR lines are observed, which are indicative of restricted rotational motion (Sect. 2.6.3). In comparison, an aqueous solution of TEMPO at this temperature exhibits a three-line spectrum close to the fast limit. Since TEMPO is a small molecule structure analog of the nitroxide in the polymeric system, the restricted rotational mobility can be attributed to the covalent attachment to the polymer backbone. In fact, spectral simulations reveal an increase of the rotational correlation time by roughly three orders of magnitude.

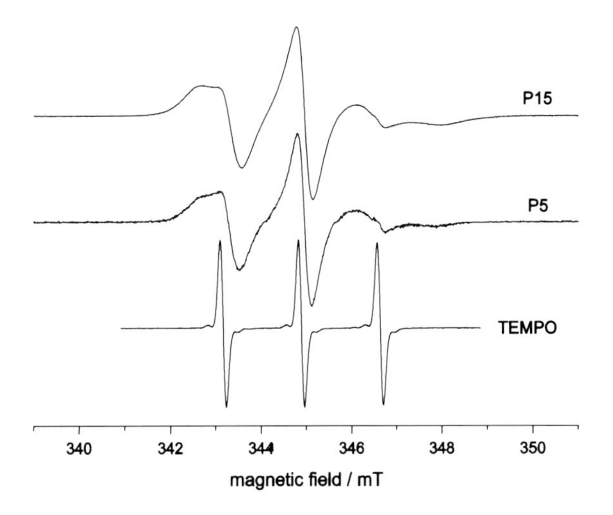

Fig. 6.1 CW EPR spectra of the EMAAm based hydrogels P5 and P15 at 278 K. The spectrum of 0.2 mM TEMPO at the same temperature is given for comparison [54]—Reproduced by Permission of the PCCP Owner Societies

This hindered rotation cannot be described by an isotropic Brownian motion. Rather, an anisotropy along and perpendicular to the labeling axis has to be introduced to obtain a good agreement of spectrum and simulation. This anisotropy is commonly observed for nitroxides which are chemically attached to a macromolecular system [46]. It manifests itself in additional features in the high- and low-field region of the spectrum. These additional peaks decrease when the temperature increases due to an increase of the rotational mobility (data not shown).

Further, it was found that water, which was expelled from a spin-labeled hydrogel during the collapse, does not give rise to an EPR signal. This is further evidence that the radicals are spatially restricted to the polymeric system and that a quantitative separation of the radicals from the solution is achieved.

6.3.3 Characteristic DNP Factors and ^{1}H Relaxation Times

The maximum achievable NMR signal enhancement is a crucial factor for a DNP polarization agent. In Fig. 6.2a, the highest observed ^{1}H-DNP enhanced NMR signal of water is shown. NMR signal enhancements of up to $E = -21.2 \pm 1.1$ (P5) and $E = -26.6 \pm 1.3$ (P15) were achieved at a microwave (mw) power of 2 W. In comparison, a solution of free TEMPOL (c = 10 mM) gives rise to ^{1}H signal enhancement of $E = -148.8$ at the same mw power (15 °C).

For mw powers exceeding 2 W, a decline of the enhancement was observed for all adjusted temperatures (Fig. 6.2b). This decline is rationalized by an irradiation induced fast thermal collapse (<1 s) of the hydrogel due to heating of the water above T_C even in a cooled system. Thus the incorporated water is expelled and locally separated from the spin-labeled hydrogel. This leads to a diminished

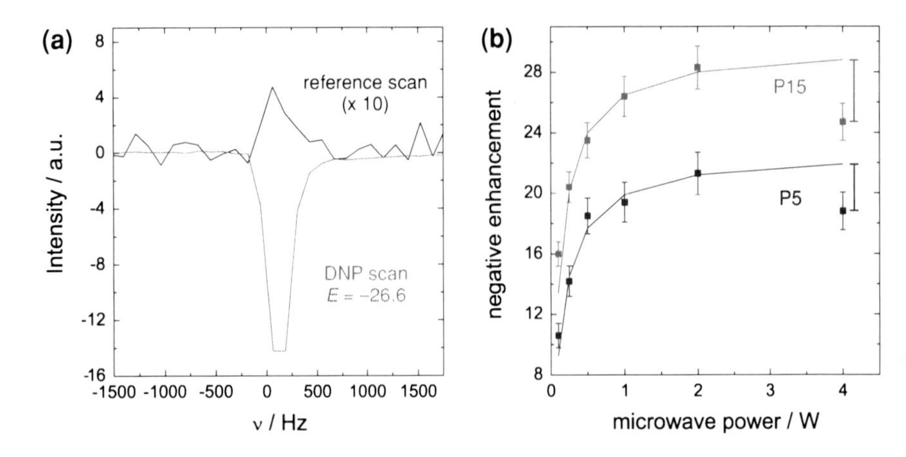

Fig. 6.2 **a** DNP enhanced (*red*) and reference (*black*) ^{1}H NMR spectra of water-swollen P15 at $T = 5$ °C and a microwave power of 2 W. **b** Plot of the enhancement against the microwave power for P5 and P15. In both cases, a decrease of the enhancement at 4 W is observed, which is caused by a partial mw induced collapse of the hydrogel. The discrepancy between the expected and the actually measured signal enhancement at 4 W is indicated by bars [54]—Reproduced by Permission of the PCCP Owner Societies

coupling factor before the sample is completely saturated and, consequently, to an inefficient hyperpolarization. This conclusion is supported by DNP experiments with NiPAAm based spin labels, which show a significantly decreased enhancement up to only -16.2 (data not shown).

It is likely, though, that the samples are only partially collapsed due to local heating. However, as explained in the last chapter, even at temperatures below T_C a microphase separation takes place. This, in turn, explains the reduced enhancement for high mw powers, which is slightly smaller than the reported value by McCarney et al. for ^{14}N spin-labeled sepharose gels ($E = -33$ at $6\,\mathrm{W}, c = 10\,\mathrm{mM}$) [31].

Interestingly, the proton spin-lattice relaxation times $T_{1,0}$ in the hydrogel network are extremely short even without spin labels ($\sim$480 ms) as compared to the relaxation time of free water ($\sim$2,100 ms). This behavior was observed before and was attributed to the sterical confinement in the hydrogel network, since a correlation to the pore size of the network was found [21]. Small changes of the swelling degree lead to a marked change of the spin-lattice relaxation times as the degree of swelling severely affects the pore size and thus the spin-lattice relaxation time. Thus, uncertainties of the T_1 measurements were estimated as large as 20%.

The relaxation times were further decreased to $T_1 \sim 330$ ms (P5) and $T_1 \sim 65$ ms (P15) in the presence of spin labels. Thus, the leakage factor was determined as $f = 0.31 \pm 0.20$ (P5) and $f = 0.86 \pm 0.04$ (P15). The low leakage factor for P5 illustrates the impact of the reduced proton spin lattice relaxation $T_{1,0}$ on the DNP efficiency. Due to the severe reduction caused by the hydrogel network, the spin

Table 6.1 DNP parameters for P5 at 5 °C and P15 at 15 °C

	c(SL)/mM	E_{max}	E	f	ξs
P5	1.0 ± 0.2	–22.9 ± 1.2	–21.2 ± 1.1	0.31 ± 0.20	0.108 ± 0.068
P15	6.8 ± 0.5	–29.5 ± 1.5	–26.6 ± 1.3	0.86 ± 0.04	0.049 ± 0.004

labels are less effective polarizing agents. The leakage factor f can be improved effectively only with an increased labeling degree, as observed by the elevated leakage factor of P15.

Due to the large error estimate for f and difficulties to separate the saturation s and the coupling factor ξ for the studied hydrogel systems, the product ξs is reported. The observed and calculated DNP parameters are summarized in Table 6.1 for the best achieved enhancements of the spin-labeled hydrogels.

Since ξ is independent of the labeling degree, P5 is easier to saturate. This conclusion is supported by a line width analysis of the EPR center line, $\Delta m_{\mathrm{I}} = 0$. A narrower line width (FWHM ~13.4 MHz) is observed for P5 compared to P15 (FWHM ~17.2 MHz), which, in principle, allows to saturate more electron spin packets and to achieve a higher DNP enhancement. Hence, the higher enhancement in P15 solely results from its higher leakage factor. The difficulties in saturation also occur because (i) the radicals are immobilized in the network, so that Heisenberg spin exchange is suppressed and (ii) the mixing of the EPR lines by nuclear spin relaxation cannot be exploited due to power limitations caused by the collapse of hydrogel network.

Although the thermal collapse of the studied systems limits the achievable DNP, it opens the possibility to combine the saturation of the EPR lines and the separation of hyperpolarized target molecules and radicals in a single step. In addition, the spatial separation from the radicals provides for a significantly prolonged spin-lattice relaxation time of the hyperpolarized molecules. A relaxation time of T_1 ~2100 ms was measured for water expelled from the hydrogel at 25 °C.

6.3.4 Temperature-Dependent DNP Performance of Spin-Labeled Hydrogels

In Fig. 6.3, the measured DNP enhancements are plotted against the adjusted sample temperature. In principle, a temperature elevation should result in a higher coupling factor and in an increase of the signal enhancement. However, the highest enhancements were achieved at low temperatures (5, 15 and 25 °C) when high mw powers were applied. At elevated temperatures, i.e. 45 and 55 °C, a substantial decrease of the enhancement is observed. This decrease results from a partially collapsed hydrogel network, which causes incomplete saturation and small dipolar coupling between electron and proton spins, as explained in the last section.

In contrast to the sharp macroscopic collapse, a smooth decrease of the DNP enhancements is observed. This continuous decrease can be rationalized bearing in

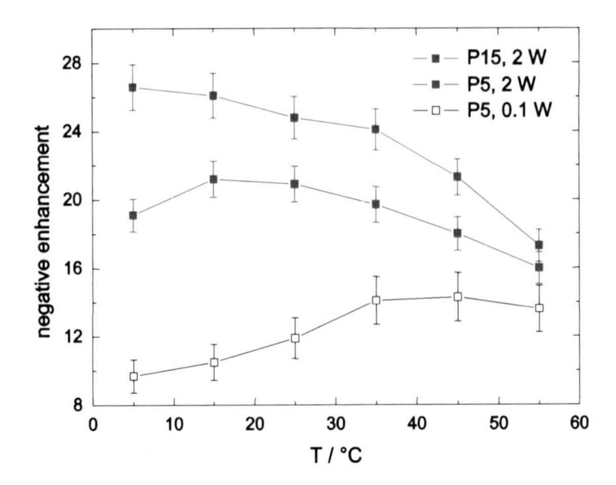

Fig. 6.3 Temperature dependence of the ^{1}H NMR signal enhancement at high mw power (2 W, *red* and *blue* line) and low power (0.1 W, *black* line). While low power DNP experiments show an expected increase of the signal enhancement up to 45 °C, a decrease is observed when applying high mw power. This decrease is caused by mw heating, which induces a partial local collapse of the hydrogel [54]—Reproduced by Permission of the PCCP Owner Societies

mind that the polarization transfer is governed by radical–water interactions. Hence, the nanoscopic structure of the hydrogel rather than its macroscopic topology is responsible for these interactions. As shown in the last chapter, locally collapsed regions are continuously built up over a broad temperature range. These collapsed regions will not efficiently contribute to a DNP effect, since they are in the vicinity of only few, strongly immobilized water molecules. Thus, the continuous formation of these collapsed regions is described by the smooth reduction of the DNP enhancement. In fact, it contains the same information about the hydrogel nanostructure, however, not observed from the radical's point of view but from the position of a water molecule. Similarly, Han et al. [22, 47] used the magnitude of the DNP signal enhancement to characterize the local structure of and water transport through lipid membranes.

By reducing the mw power to 0.1 W—which is not enough to substantially heat the sample during mw irradiation—the intrinsic DNP temperature dependence from the heating-induced collapse effect could be separated. Under these experimental conditions, the expected rise of the DNP enhancement with temperature was found until the temperature approaches T_C (Fig. 6.3). To find the best enhancement factor, one has to compromise between two opposing effects. First, the 'intrinsic' coupling factor increases with higher temperature due to the higher spin label rotational mobility, which results in slightly narrowed EPR lines. Second, the mw heating-induced collapse of the hydrogel network separates the spin labels from water and potentially polarizable biomolecules resulting in lower DNP enhancements.

6.4 Conclusions

In summary, the spin-labeled hydrogels introduced here are promising polarizing agents for in situ and ex situ DNP experiments. Specifically, the hyperpolarized systems benefit from a prolonged T_1 relaxation time since radicals and polarized molecules are effectively separated due to the macroscopic collapse. For in situ DNP methods, the prolonged T_1 time and the suppressed radical-induced line broadening may allow for more complex 2D NMR experiments which might be beneficial for the study of biomolecules (e.g. proteins). For ex situ DNP methods, the prolonged lifetime of hyperpolarization reduces the polarization loss during the transport of the hyperpolarized molecules. Additionally, the obtained hyperpolarized biomolecules in the radical-free, non-toxic solute can be utilized for biomedical applications.

The biggest advantage of the spin-labeled hydrogels, the thermally induced collapse, is also their biggest disadvantage. The mw induced heating of the gel causes a partial collapse, which in turn decreases the DNP efficiency. This prevents the use of high mw powers to effectively saturate the EPR transitions and to achieve high enhancements. This limitation can be overcome by dissolution DNP at low temperatures (a few K), since a microwave-induced heating will not result in a structural variation of the gel. The application of spin-labeled gels in dissolution DNP and strategies for improved polarization efficiencies are discussed in the next section.

6.5 Outlook

In dissolution DNP, the substantially increased Boltzmann polarization of electron spins at low temperatures is transferred to nuclear spins to achieve enhancement factors $>10{,}000$ compared to the equilibrium polarization at room temperature [28]. After the polarization, the sample is ramp-heated from its ultracold state to ambient temperature and shuttled to the NMR detection magnet. Since the polarization takes place at very low temperatures, severe microwave heating effects of the gel are overcome. Nonetheless, the gel still exhibits its biggest strength, as it still provides for an efficient separation of radicals and solute during the subsequent heating process.

While polarization in the liquid state is governed by the Overhauser effect, two mechanisms are mainly responsible for a polarization transfer in solids, the cross effect and thermal mixing [48, 49]. These mechanisms are based on three-spin processes involving two electron spins and one nuclear spin. They are most efficient if the dipolar coupling of the electron spins matches the Larmor frequency of the respective nucleus. This is achieved by a specific distance and $\mathbf{g}$ tensor orientation of the two radicals. Thus, biradicals with optimized molecular structures lead to significantly increased DNP enhancements [25, 50, 51].

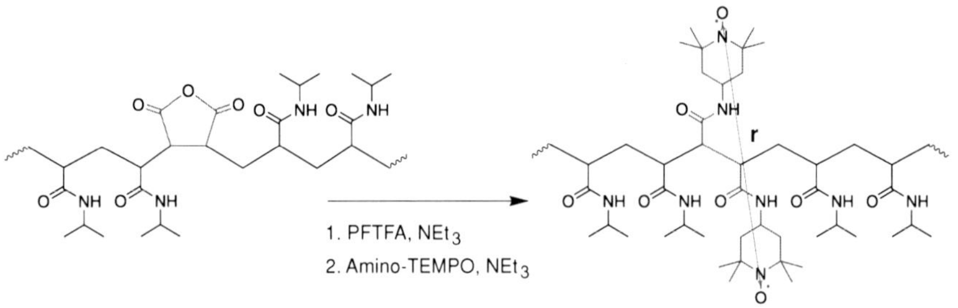

1. PFTFA, NEt_3
2. Amino-TEMPO, NEt_3

Scheme 6.4 Proposed reaction scheme leading to a biradical analog unit in a thermoresponsive hydrogel based on *N*-isopropylacrylamide. Via force field calculations, a distance of 1.47 nm was found between the two electrons, which were assumed to be located in the center of the N–O bond

To optimize the spin-labeled responsive hydrogels with respect to their solid-state DNP properties, the incorporation of a biradical analog structure is favorable. This can be achieved by a copolymerization of a thermoresponsive compound and a crosslinker with maleic anhydride instead of methacrylic acid. The anhydride units can be then transformed to two adjacent radicals, as illustrated in Scheme 6.4. In this scheme, NiPAAm is depicted as thermoresponsive unit since dissolution DNP does not impose the requirement of a high critical temperature.

Preliminary force field calculations (MMFF94, Chem3D Ultra) on a polymer chain fragment suggested that the adjacent radical units assume a *trans* state with an electron–electron distance of $r = 1.47$ nm. This is reasonable due to the sterical hindrance of these bulky side groups. Assuming $g_{N,iso} = 2.0060$ and utilizing Eq. 2.101, this distance gives rise to a dipolar coupling between the electrons of $\nu_{\perp} = 16.44$ MHz, which is close to the proton Larmor frequency at X-band [$\nu(^1H) = 14.9$ MHz at 350 mT]. For an optimal DNP transfer, the piperidine rings have to be tilted by 90° with respect to each other to provide for an ideal orientation of the **g** tensor. This orientational restriction is not achieved with the synthetic approach presented here. However, the system experiences a large degree of freedom and will assume a range of orientations with the desired orientation among them.

6.6 Materials and Methods

Materials. Methacrylic acid (MAA, 99%, Aldrich) was distilled prior to use. *N,N*-Ethylmethylacrylamide (EMAAm) and 4-methacryloyloxybenzophenone (MABP) were prepared according to reported procedures [52, 53]. Dioxane was distilled over CaH_2 and dried over a molecular sieve (4 Å). 2,2′-Azobis(isobutyronitrile) (AIBN, 98%, Acros) was recrystallized from methanol. Triethylamine (NEt_3, 99.5%, Fluka), pentafluorophenyltrifluoroacetate (PFTFA, 98%, Aldrich)

Table 6.2 Properties of the synthesized polymers

	Monomer composition			M_w/ g mol^{-1} [a]	M_w/ M_n^a	Yield / %
	EMAAm	MAA	MABP			
P5	94	5	1	49,600	3.1	89
P15	84	15	1	28,000	4.2	92

[a] determined by GPC at 60 °C in DMF, standard: PMMA

and 4-amino-2,2,6,6-tetramethylpiperidine-1-oxyl (Amino-TEMPO, 97%, Acros) were used as received. Distilled water was further purified by a MilliQ System (Millipore) to achieve a resistivity of 18.2 MΩ cm.

Thermoresponsive Polymer. Thermoresponsive terpolymers based on EMA-Am were synthesized by Michelle Drechsler according to the reaction scheme introduced in Sect. 5.5. After purification, they were freeze-dried from water in vacuo. The yield was around 90%. ^{1}H-NMR (250 MHz, d^4-MeOH): δ/ppm = 0.9–1.2 (m, -**CH$_3$**, MAA, MABP, -CH$_2$**CH$_3$** EMAAm), 1.5–1.9 (m, **CH$_2$** backbone), 2.2–3.7 (m, **CH** backbone, -CH$_2$**CH$_3$**, -**Me**, EMAAm), 6.8–8.0 (m, C-**H$_{arom}$**). The molecular weight and the polydispersity index (M_w/M_n) were determined by gel permeation chromatography with dimethylformamide as mobile phase. The measurement was conducted at 60 °C, PMMA served as internal standard. The monomer composition and the molecular weight distribution of the synthesized polymers are summarized in Table 6.2. The suffixes 5 and 15 denote polymers synthesized with 5 and 15% MAA in the monomer feed, respectively.

Spin Labeling and Sample Preparation. 15 wt% polymer solutions in ethanol were drop-cast on hexamethyldisilazane modified glass slides and dried over night in vacuo at 40 °C to achieve polymeric films in the range of 10–15 μm. The polymer was crosslinked and tethered to the substrate by UV irradiation (6.82 J/cm^2) with a Stratagene UV Stratalinker with a peak wavelength of 365 nm. The carboxylic acid groups of the resulting gels were converted to active ester units with a two fold excess of PFTFA and NEt_3 in dichloromethane (DCM). After washing the gels with DCM twice, they were reacted with a two fold excess of 4-amino-TEMPO and NEt_3 in DCM to achieve the spin-labeled material. The synthetic route is described in Scheme 6.5. In order to remove unreacted nitroxide molecules, the gels were subjected to dialysis (Spectra/Por 3 Dialysis Membrane, MWCO = 3500 g mol^{-1}) in ethanol for one week. The solvent was removed in vacuo and the spin-labeled gel was abraded with a scalpel, filled into EPR sample canules (i.d. 0.8 mm) and swollen with deionized water. Excess water not bound to the hydrogel was removed.

Experimental Details. All DNP and EPR measurements were performed by Björn Dollmann on a Bruker Elexsys 580 spectrometer at an external magnetic field of 0.345 T, corresponding to a proton Larmor frequency of $\nu(^1H) = 14.7$ MHz. CW EPR spectra were detected with a critically coupled Bruker EN4118K-MD4 ENDOR probe head. For the NMR detection, the ENDOR probe was externally tuned and matched to the operating NMR frequency. The DNP effect was measured in dependence of the irradiated mw power and the

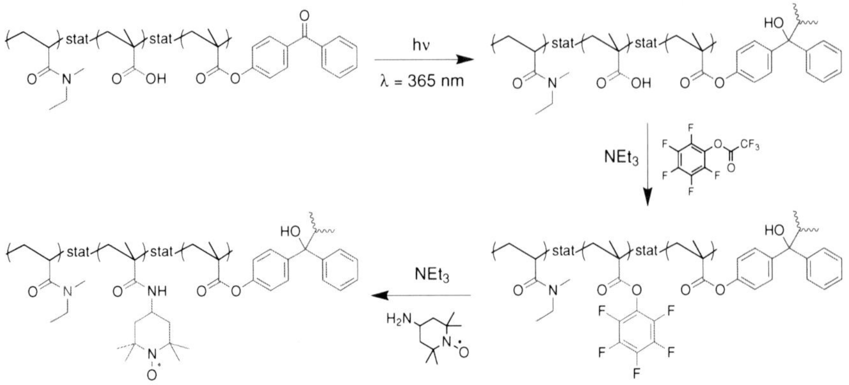

Scheme 6.5 Synthetic route to the spin-labeled hydrogel via active ester formation [54]—Reproduced by Permission of the PCCP Owner Societies

temperature in the range of 5–55 °C (cf. $T_C \sim 63$°C). CW EPR spectra were recorded for all temperatures. The temperature was adjusted with a closed cycle cryostat (ARS AF204, customized for pulse EPR, ARS, Macungie, PA, USA). The continuous mw irradiation time on the EPR center line was kept as short as possible (ranging from $3T_1$ of water in hydrogels to ≤ 1 s) to minimize heating effects. The spin label concentration of the hydrogels in the swollen state was determined by CW EPR calibration measurements with 4-hydroxy-2,2,6,6-tetramethylpiperidine-1-oxyl (TEMPOL) solutions of known concentration. Spin label concentrations of ~ 1 and ~ 6.8 mM were found for P5 and P15.

References

1. Bargon J, Fischer H, Johnsen U (1967) Z Naturforsch A 22:1551–1555
2. Bargon J, Fischer H (1967) Z Naturforsch A 22:1556–1562
3. Hore PJ, Broadhurst RW (1993) Prog Nucl Magn Reson Spectrosc 25:345–402
4. Matysik J, Diller A, Roy E, Alia A (2009) Photosynth Res 102:427–435
5. Bowers CR, Weitekamp DP (1986) Phys Rev Lett 57:2645–2648
6. Bowers CR, Weitekamp DP (1987) J Am Chem Soc 109:5541–5542
7. Eisenschmid TC, Kirss RU, Deutsch PP, Hommeltoft SI, Eisenberg R, Bargon J, Lawler RG, Balch AL (1987) J Am Chem Soc 109:8089–8091
8. Pravica MG, Weitekamp DP (1988) Chem Phys Lett 145:255–258
9. Goodson BM (2002) J Magn Reson 155:157–216
10. Overhauser AW (1953) Phys Rev 92:411–415
11. Carver TR, Slichter CP (1953) Phys Rev 92:212–213
12. Carver TR, Slichter CP (1956) Phys Rev 102:975–980
13. Becerra LR, Gerfen GJ, Temkin RJ, Singel DJ, Griffin RG (1993) Phys Rev Lett 71:3561–3564
14. Bowen S, Hilty C (2008) Angew Chem Int Ed 47:5235–5237

15. Bajaj VS, Mak-Jurkauskas ML, Belenky M, Herzfeld J, Griffin RG (2009) Proc Nat Acad Sci USA 106:9244–9249
16. Day SE, Kettunen MI, Gallagher FA, Hu DE, Lerche M, Wolber J, Golman K, Ardenkjaer-Larsen JH, Brindle KM (2007) Nat Med 13:1521
17. Golman K, in't Zandt R, Lerche M, Pehrson R, Ardenkjaer-Larsen JH (2006) Cancer Res 66:10855–10860
18. Gallagher FA, Kettunen MI, Day SE, Hu DE, Ardenkjaer-Larsen JH, in't Zandt R, Jensen PR, Karlsson M, Golman K, Lerche MH, Brindle KM (2008) Nature 453:940–943
19. Prisner T, Köckenberger W (2008) Appl Magn Reson 34:213–218
20. Sezer D, Gafurov M, Prandolini MJ, Denysenkov VP, Prisner TF (2009) Phys Chem Chem Phys 11:6638–6653
21. McCarney ER, Armstrong BD, Lingwood MD, Han S (2007) Proc Nat Acad Sci USA 104:1754–1759
22. McCarney ER, Armstrong BD, Kausik R, Han S (2008) Langmuir 24:10062–10072
23. Armstrong BD, Lingwood MD, McCarney ER, Brown ER, Blümler P, Han S (2008) J Magn Reson 191:273–281
24. Münnemann K, Bauer C, Schmiedeskamp J, Spiess HW, Schreiber WG, Hinderberger D (2008) Appl Magn Reson 34:321–330
25. Song C, Hu K-N, Joo C-G, Swager TM, Griffin RG (2006) J Am Chem Soc 128: 11385–11390
26. Maly T, Debelouchina GT, Bajaj VS, Hu K-N, Joo C-G, Mak-Jurkauskas ML, Sirigiri JR, van der Wel PCA, Herzfeld J, Temkin RJ, Griffin RG (2008) J Chem Phys 128:052211
27. Höfer P, Parigi G, Luchinat C, Carl P, Guthausen G, Reese M, Carlomagno T, Griesinger C, Bennati M (2008) J Am Chem Soc 130:3254–3255
28. Ardenkjaer-Larsen JH, Fridlund B, Gram A, Hansson G, Hansson L, Lerche MH, Servin R, Thaning M, Golman K (2003) Proc Nat Acad Sci USA 100:10158–10163
29. Comment A, Rentsch J, Kurdzesau F, Jannin S, Uffmann K, van Heeswijk RB, Hautle P, Konter JA, van den Brandt B, van der Klink JJ (2008) J Magn Reson 194:152–155
30. Gitti R, Wild C, Tsiao C, Zimmer K, Glass TE, Dorn HC (1988) J Am Chem Soc 110: 2294–2296
31. McCarney ER, Han S (2008) J Magn Reson 190:307–315
32. Thaning M, Servin R (GE Healthcare, Inc.) Method of dynamic nuclear polarisation DNP, USPTO patent application 20080260649, applied on 12-1-2006, published on 10-23-2008
33. Hirotsu S, Hirokawa Y, Tanaka T (1987) J Chem Phys 87:1392–1395
34. Schild HG (1992) Prog Polym Sci 17:163–249
35. Hausser KH, Stehlik D (1968) Adv Magn Reson9 3:79–139
36. Nicholson I, Lurie DJ, Robb FJL (1994) J Magn Reson B 104:250–255
37. Benial AMF, Ichikawa K, Murugesan R, Yamada K, Utsumi H (2006) J Magn Reson 182:273–282
38. Bates RD, Drozdoski WS (1977) J Chem Phys 67:4038–4044
39. Robinson BH, Haas DA, Mailer C (1994) Science 263:490–493
40. Armstrong BD, Han S (2007) J Chem Phys 127:104508
41. Beines PW, Klosterkamp I, Menges B, Jonas U, Knoll W (2007) Langmuir 23:2231–2238
42. Gianneli M, Beines PW, Roskamp RF, Koynov K, Fytas G, Knoll W (2007) J Phys Chem C 111:13205–13211
43. Junk MJN, Jonas U, Hinderberger D (2008) Small 4:1485–1493
44. Cao Y, Zhu XX, Luo J, Liu H (2007) Macromolecules 40:6481–6488
45. Cao Y, Zhao N, Wu K, Zhu XX (2009) Langmuir 25:1699–1704
46. Atherton NM (1993) Principles of electron spin resonance. Ellis Horwood, New York
47. Han S, McCarney ER, Armstrong BD (2008) Appl Magn Reson 34:439–451
48. Wind RA, Duijvestijn MJ, van der Lugt C, Manenschijn A, Vriend J (1985) Prog Nucl Magn Reson Spectrosc 17:33–67
49. Abragam A, Goldman M (1978) Rep Prog Phys 41:395–467
50. Hu K-N, Yu H-h, Swager TM, Griffin RG (2004) J Am Chem Soc 126:10844–10845

51. Hu K-N, Song C, Yu H-h, Swager TM, Griffin RG (2008) J Chem Phys 128:052302
52. Shea KJ, Stoddard GJ, Shavelle DM, Wakui F, Choate RM (1990) Macromolecules 23: 4497–4507
53. Toomey R, Freidank D, Rühe J (2004) Macromolecules 37:882–887
54. Dollmann BC, Junk MJN, Drechsler M, Spiess HW, Hinderberger D, Münnemann K (2010) Phys Chem Chem Phys 12:5879–5882

Chapter 7
Local Nanoscopic Heterogeneities in Thermoresponsive Dendronized Polymers

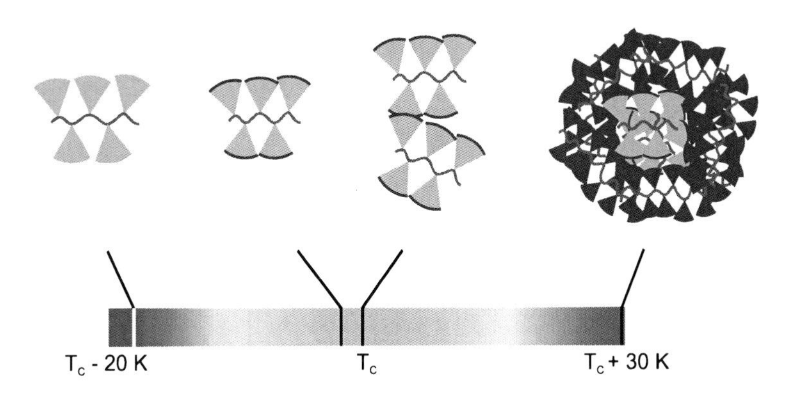

The thermal transition of thermoresponsive dendronized polymers is characterized on a molecular scale by continuous wave EPR spectroscopy. It is found to be accompanied by dynamic structural heterogeneities on the nanoscale, which trigger the aggregation of single polymer chains into mesoglobules. While macroscopically a sharp phase transition, this study reveals that the dehydration of the polymer chains proceeds over a temperature interval of at least 30 K and is a case of a molecularly controlled non-equilibrium state. While the aggregation temperature mainly depends on the periphery of the dendrons, the dehydration of the mesoglobule is governed by the hydrophobicity of the dendritic core. Heating rate dependent changes were assigned to the formation of a dense polymeric layer at the periphery of the mesoglobule, which prohibits the release of incorporated water.

M. J. N. Junk, *Assessing the Functional Structure of Molecular Transporters by EPR Spectroscopy*, Springer Theses, DOI: 10.1007/978-3-642-25135-1_7,

7.1 Introduction

As already stated in the last chapters, thermoresponsive polymeric materials are of great interest for their potential use in various fields including actuators, (targeted) drug delivery and surface modification [1–7]. Ever since Wu's discovery of the coil–globule transition of single poly(*N*-isopropylacrylamid) (pNiPAAm) chains near the lower critical solution temperature (LCST) [8–11], the collapse mechanism including the formation of stable mesoglobules has been an intensely studied topic [12–18]. Despite these efforts, a molecular scale picture of what happens when thermoresponsive polymers start to dehydrate at a certain temperature, subsequently collapse and self-assemble into mesoglobules, does not exist. This severely hampers rational materials design.

In an exploratory research effort aiming at detecting unusual properties of dendronized polymers [19–23], it was recently found that such systems based on oligoethyleneglycole (OEG) units exhibit fast and fully reversible phase transitions with a sharpness that is amongst the most extreme ever observed [24–26]. These dendronized polymers are soluble in water and their lower critical solution temperature (LCST) is found in a physiologically interesting temperature regime between 30 and 36 °C, which is as low as it is known for poly(ethylene oxide) and long chain ethylene oxide oligomers. For the latter, the influence of hydrophobic end groups on the LCST has been thoroughly investigated both experimentally and theoretically as summarized in a recent review [27].

Given this extraordinary behavior, these polymers may be well suited to gain a deeper understanding of the processes involved. There are indications that the thermal transition proceeds via the formation of structural inhomogeneities of variable lifetimes on the nanometer scale that are still poorly understood. Indeed, this topic has been identified as one of the major challenges of research in the macromolecular sciences in the coming years [28]. In particular, this concerns a clearer understanding of the formation, structure, and lifetimes of these local inhomogeneities, the effect of the individual chemical structures on the physical processes, and the consequences that the local heterogeneities bear for the aspired function (e.g. drug delivery). To obtain insight into structure–function relations it is essential to develop and use characterization techniques capable of resolving structural heterogeneities on length scales between sub-nanometers and several tens of nanometers and temporal fluctuations in a window of sub-nano- to microseconds.

The remarkable macroscopic behavior results from the systems being far from classical macroscopic equilibrium. It can be viewed as an example of 'molecularly controlled non-equilibrium'. We note that macromolecule-based processes far from equilibrium are extensively found in nature, e.g. in DNA replication, to obtain high specificity in the noisy environment of a cell [29, 30]. However, investigations into similar concepts in synthetic macromolecular systems are still rare [31].

Magnetic resonance techniques as intrinsically local methods meet the conditions required to solve questions involved with structural inhomogeneities of functional macromolecules [32–34] and dynamic heterogeneities in polymer melts in the vicinity of the glass transition [35–37]. The structure and dynamics of synthetic as well as biological macromolecules on the molecular scale were extensively studied by advanced NMR and EPR techniques [31–44]. A particularly simple way of studying the molecular environment of dendronized polymers, which undergo a thermal transition, utilizes conventional continuous wave (CW) EPR spectroscopy on nitroxide radicals as paramagnetic tracer molecules. As described in detail in Sects. 2.6.3 and 2.6.4 and utilized in Chap. 5, these spin probes are sensitive to the local viscosity, which gives rise to changes of the rotational correlation time, and to the local polarity/hydrophilicity [34, 42–44]. The latter affects the electronic structure of the radical and changes the spectral parameters, specifically the g-factor and the hyperfine coupling constant to ^{14}N.

The amphiphilic radical 2,2,6,6-tetramethylpiperidine-1-oxyl (TEMPO, Scheme 7.1) is especially suited to sample both hydrophobic and hydrophilic regions. In Chap. 5, it was successfully applied to observe structural nano-inhomogeneities in NiPAAm-based hydrogels, which were static over a time scale of at least 2 h. Based on this experience, the same spin probe was applied to aqueous solutions of the dendronized polymers to gain insight into the molecular processes associated with the thermal transition.

The polymers in this study are polymethacrylate derivatives with first, second, and third generation triethyleneoxide dendrons, PG1–3(ET). In addition, a second generation dendronized polymethacrylate was studied, where the triethyleneoxide core was replaced by a hydrophobic octane unit (PG2(ETalkyl), Scheme 7.1). While the periphery of the dendrons is not varied, chemical modifications and different generations provide dendron cores, which substantially differ in their dimension and in their hydrophilicity.

This chapter is organized as follows. In the first part, the results of the spin probe study above the critical temperature T_C are described and discussed. Special emphasis is placed on what happens on the nanoscopic scale as the polymer solutions phase separate from water and form mesoglobules. The second part focuses on conformational changes of the polymers below T_C, i.e. before aggregation of the polymer takes place. In the third part, the effect of the heating rate on the mesoglobule formation in general and on the distribution of spin probes in particular is discussed. The formation of a dense polymeric layer at the periphery of the mesoglobule is claimed responsible for the observed changes. The proposed model is discussed in light of the obtained EPR spectroscopic results and other reported data on the mesoglobule formation. Finally, the results are summarized and concluded with a special emphasis on the influence that the dendritic cores exert on the observed temperature-induced processes.

7.2 Local Heterogeneities Above the Critical Temperature

Representative CW EPR spectra of TEMPO in an aqueous solution of 10 wt% PG1(ET) well above and below the critical temperature T_C of 33 °C are shown in Fig. 7.1a. While the low field and the center peaks remain almost unaffected, the

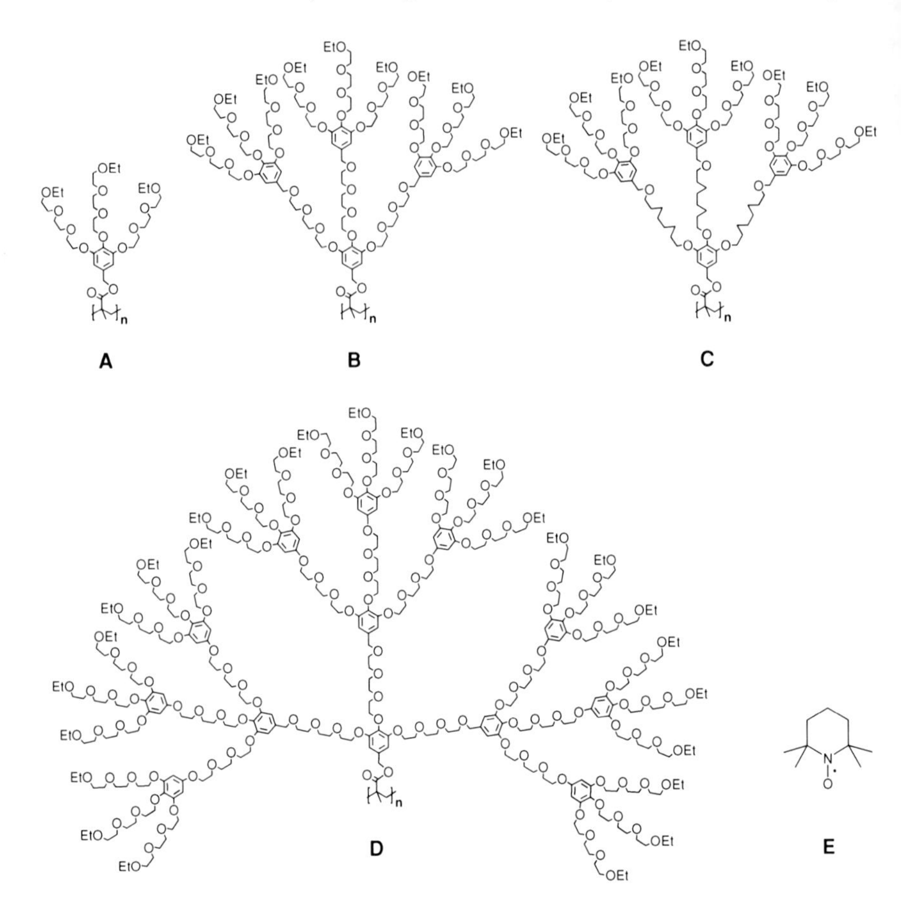

Scheme 7.1 Chemical structures of the thermoresponsive dendronized polymers **a** PG1(ET), **b** PG2(ET), **c** PG2(ETalkyl), **d** PG3(ET), and **e** the spin probe TEMPO

high field line, most sensitive to structural and dynamic effects, changes considerably and is displayed in Fig. 7.1b for various temperatures. The apparent splitting of this line at elevated temperatures originates from two nitroxide species A and B that are placed in local environments with different polarities (cf. Chap. 5). This gives rise to considerable differences of the isotropic hyperfine coupling constants a_{iso} and g-values g_{iso}.

Before proceeding further, it is checked whether the critical temperatures from turbidity measurements and EPR spectroscopy coincide. Turbidity measurements reflect a phase separation process of a seemingly classical nature where droplets of a concentrated solution of the polymer separate from the dilute solution of the polymer (binodal decomposition). These droplets of the concentrated phase were identified by light microscopy [24]. The critical temperature T_{C} derived from EPR is defined as the first pronounced decrease of the hydrophilic fraction y_{A} (Appendix A.4.2). This temperature coincides with the lowest temperature,

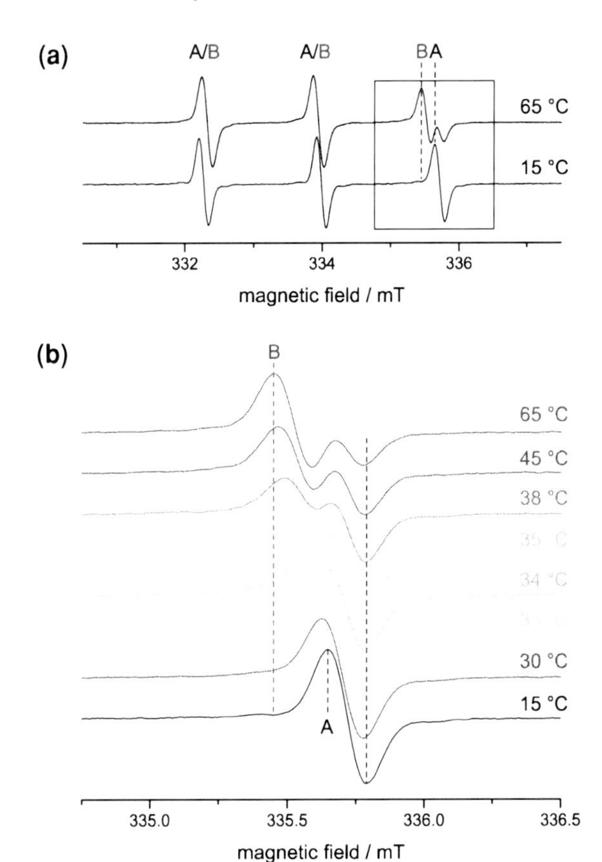

Fig. 7.1 **a** CW EPR spectra of 0.2 mM TEMPO in an aqueous solution of 10 wt% PG1(ET) recorded at 15 and 65 °C and **b** detailed plot of the high field transition $m_I = -1$, which is marked by a rectangle in **a** at selected temperatures. The contribution to the high field peak at 335.4 mT, denoted B, originates from TEMPO molecules in a hydrophobic environment, while nitroxides in a hydrophilic surrounding give rise to the contribution A. The broken lines at the outer extrema of the peak serve as guides to the eye. Reprinted with permission from [73, 74]. Copyright 2010 WILEY-VCH Verlag GmbH & Co. KGaA, Weinheim

at which two separate contributions to the EPR signal can be recognized. Though the obtained EPR curves are very broad and extent to temperatures far beyond T_C, this first drop is well-defined ($\Delta T_C = 1$ K).

Indeed, the critical temperatures for PG1(ET) $T_C = 32$ °C (turbidity: 33 °C), PG2(ET) 34 °C (36 °C), PG3(ET) 34 °C (34 °C), and PG2(ETalkyl) 30 °C (31 °C) as obtained from EPR and turbidity measurements are almost identical. The slightly lower EPR-derived values are due to higher concentrations of the polymer solutions. The macroscopic phase separation identified by turbidimetry is a consequence of the well-known fact that the solvent quality of water with respect to oxyethylene segments is strongly temperature dependent. In other words, water becomes a thermodynamically poor solvent as the temperature increases. Yet, the gel phase formed in equilibrium with the dilute phase is still highly swollen by water. The properties of this gel phase as seen by the spin probe form the objective of the studies presented below.

Similar to the swollen pNiPAAm network in Chap. 5, the spectral parameters for component A coincide with those of TEMPO in pure water ($a_A \sim 48.3$ MHz), i.e. this spin probe is located in a strongly hydrated, hydrophilic environment. The observed decrease of a_{iso} by 3.7 MHz for species B (at 65 °C) is indicative of much more hydrophobic and less hydrated surroundings for these spin probes (comparable to chloroform or *tert*-butylalcohol) [45]. At temperatures below T_{C}, only the hydrophilic spectral component A is observed since all dendritic units are water–swollen (cf. Sect. 7.3). Above the critical temperature of 33 °C, an increasing fraction of hydrophobic species B is observed with increasing temperature. The dehydration of the dendritic units thus leads to a local phase separation with the formation of hydrophobic cavities.

More strikingly and in contrast to the pNiPAAm studies in Chap. 5, the peak position of the spectral component B is not fixed but approaches its final value only at temperatures well above T_{C}. This indicates a dynamic exchange of the spin probes between hydrophilic and hydrophobic regions as explained in detail in Sect. 2.5. This exchange leads to an intermediate hyperfine coupling constant that is an effective weighted average between the two extreme values of the hydrophilic and the (static) hydrophobic regions (at 65 °C). Thus, the inhomogeneities formed upon the phase separation are not static, but dynamic and strongly influence the EPR spectral shape.

The exchange detected by the spin probes can be caused by two effects: hopping of the spin probe between collapsed and hydrated polymer aggregate regions, or fluctuations of the aggregates themselves. The latter can be viewed as fast opening and closing of hydrophobic cavities or a fast swelling and de-swelling of regions surrounding the spin probe. The size of the inhomogeneities can be estimated by the translational displacement of TEMPO in the polymer matrix, given by $\langle x^2 \rangle = 6D_{\mathrm{T}}\tau_{\mathrm{T}}$. At 34 °C, a maximum translational displacement $\langle x^2 \rangle^{1/2} \leq 5.1$ nm of the spin probes due to diffusion is obtained (details are given in Appendix A.4.3) [46, 47]. This displacement due to Brownian motion is assisted by fluctuations of the polymer undergoing the thermal transition.

Hence, one can conclude that slightly above T_{C} the few hydrophobic cavities are still small, i.e. in the range of a few nm. The spin probe movement and/or the local polymer fluctuations lead to an exchange of the probe molecules on the EPR time scale between the hydrophobic and still overwhelmingly more abundant and larger hydrophilic regions (the fraction of species A in hydrophilic regions is larger than 60%, cf. Fig. 7.2b). The spin probes thus mainly sample the interface between two fundamentally different regions. One should note that a few local dynamic heterogeneities on a nanometer scale are sufficient to induce a macroscopically observable (by turbidity measurements) transition in the sample. Remarkably, the transition is detected at the same temperature by two methods probing length scales which differ by at least two orders of magnitude. This suggests that the small aggregates detected by EPR might be visualized as crosslinks affecting the organization of the dendronized macromolecules on much larger scales. The sharp macroscopic transition can then be viewed as the onset of a complex de-swelling process that is rather broad on the molecular scale.

The existence of clusters in oligo-ethylene oxides as a function of temperature and concentration was long observed. It was attributed not only to the changes in

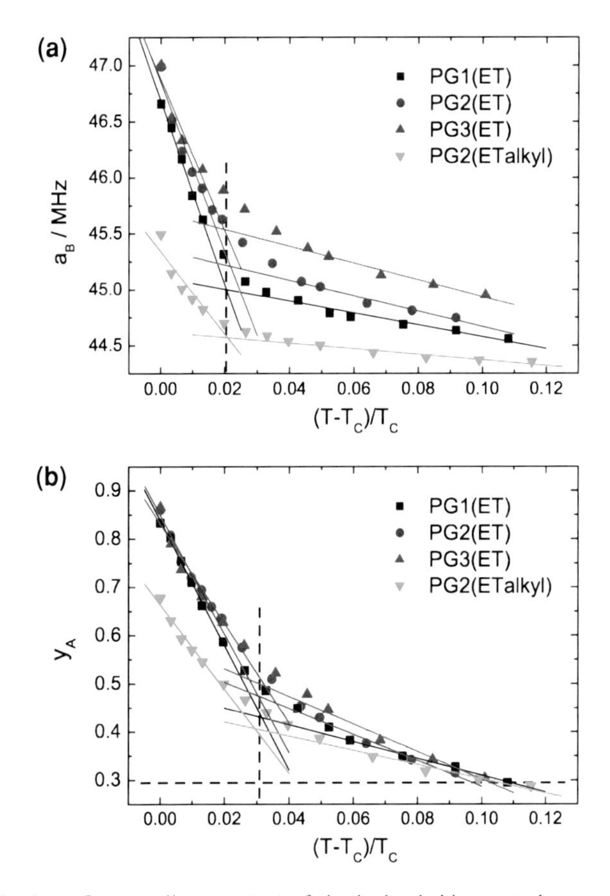

Fig. 7.2 **a** The hyperfine coupling constant of the hydrophobic spectral component a_B as a function of the reduced temperature $T_r = (T - T_C)/T_C$ for 0.2 mM TEMPO in 10 wt% aqueous solutions of four dendronized polymers, which differ in the dendron generation (PG1(ET), PG2(ET), and PG3(ET)) and the structural properties of the dendritic core (PG2(ETalkyl)). **b** Variation of the fraction of TEMPO in a hydrophilic environment y_A with increasing temperature in the above-specified polymers. Two linear fits of data points close to and far from the critical temperature illustrate at least two different de-swelling processes, clearly indicating that the temperature-induced collapse of the polymers under investigation cannot be described by a thermodynamic phase transition. The reduced temperature, at which the two lines meet, common for all polymers under investigation, is indicated by a dashed line. Reprinted with permission from [73, 74]. Copyright 2010 WILEY-VCH Verlag GmbH & Co. KGaA, Weinheim

solvent quality of water with temperature, but also to temperature-induced changes of the conformation of the oxyethylene segments. This renders them increasingly hydrophobic with increasing temperature. The latter was concluded from a temperature-dependent shift of the signals both in ^{13}C and ^{1}H NMR spectra, already below the onset of the macroscopic phase separation [48].

When increasing the temperature, not only the fraction of the hydrophobic regions but also their size grows and exchange of probe molecules between hydrophobic and hydrophilic sites becomes unlikely. The spin probes now sample the bulk hydrophobic (and remaining hydrophilic) regions rather than their interface. Together with the increase in size, the dynamics of the polymer fluctuations must slow down, as both effects are coupled. Such a slowdown in local fluctuations also adds to the observed final state of two distinct hydrophobic and hydrophilic regions that are 'static' on the EPR time scale.

To quantify the aggregation and dehydration associated with the thermally induced transition, effective hyperfine coupling constants of those TEMPO molecules in hydrophobic environments a_A and the fraction of TEMPO in hydrophilic environments y_A were determined as a function of temperature. EPR line shapes were fitted to these parameters as described in the experimental part (Sect. 7.6) and more detailed in Appendix A.4.1. By plotting these parameters vs. the reduced temperature $(T - T_C)/T_C$, it is then possible to check whether the collapse results from a well-behaved phase transition. As illustrated in Fig. 7.2, both parameters do not follow one straight line as expected for a simple phase transition [49], but instead deviate strongly from linearity. In Fig. 7.2a, static, non-exchanging hyperfine coupling values a_B are not reached until around 30 °C *above* the critical temperature. Thus, in this wide temperature range, a complex dehydration takes place which cannot be described in the picture of a classical phase transition based on a single de-swelling process.

For all polymers, at least two dehydration processes are found as indicated by the different straight lines. The first process takes place at temperatures only slightly above the critical temperature, i.e. $(T - T_C)/T_C < 0.02$, the second process in a temperature regime far above T_C. Extrapolations of the two linear fits meet at around $(T - T_C)/T_C = 0.02$ ($\sim$7 °C above T_C) for all polymers under investigation indicating that in all cases the collapse processes are equivalent. These results suggest that in the narrow temperature range up to $(T - T_C)/T_C = 0.02$, large parts of the dehydration take place, since a_B in this interval is reduced to values already close to the static final values. A further increase in the temperature results in only smaller changes of a_B, which is a sign of the expulsion of smaller amounts of residual water from the collapsed polymeric regions.

Best (straight line) fits to two processes are obtained for PG1(ET) and PG2(ETalkyl), those polymers with no or a hydrophobic dendritic core. For PG2(ET) and PG3(ET), both possessing a hydrophilic dendritic core, a significant deviation from the simple two-process fit is observed, indicating that the collapse is not fully described by two well-defined processes. Moreover, the efficiency of the first process is greatly influenced by the chemical structure of the dendritic core. It turns out to be most effective when the dehydration is supported by a hydrophobic core, as in the case of PG2(ETalkyl). It deteriorates when the core contains oxyethylene groups, which can trap more water.

Qualitatively, the same behavior is illustrated in Fig. 7.2b depicting the temperature-dependent fraction of TEMPO in a hydrophilic environment y_A. The two linear fits again intersect at the same reduced temperature for all polymers at roughly equal fractions. Further, the same dependence of the efficiency of the first strong dehydration process on the chemical structure described in Fig. 7.2a is again manifested in Fig. 7.2b. However, the graphs show one major difference. At high temperatures, y_A is only determined by the volume fraction of the collapsed polymer in water and thus approaches 0.3 for all polymer solutions. On the other hand the (static) isotropic hyperfine values of TEMPO in hydrophobic regions a_B differ depending on the structure of the dendronized polymer. PG2(ETalkyl) with an hydrophobic core provides the most hydrophobic environment, followed by PG1(ET) bearing no core. Regions with less hydrophobicity are observed for PG2(ET) with a hydrophilic ethyleneoxide core. The least hydrophobic regions are provided by PG3(ET) bearing an extended hydrophilic core. The differences cannot be explained by mere interactions of the spin probes with pure polymer chains. Rather, the hydration state provided by entrapped residual water molecules is stronger for the more hydrophilic cores and thus effectively increases the hydrophilicity of the environment of the entrapped spin probe.

The data in Fig. 7.2 support the picture of a few small hydrophobic patches triggering a macroscopically observable transition to a gel phase that is still highly swollen by water and is composed of regions differing in the local water concentration. Furthermore, the first temperature interval, marked by the steep de-hydration process, is in agreement with a growing number of uncorrelated hydrophobic regions up to a concentration and/or a volume fraction that is similar to that of the remaining hydrophilic regions (hence the kink at $y_A \sim 0.5$). This could be an indication that the growth of hydrophobic regions reaches a threshold that could be interpreted as a percolation point. When the fractions of species A to B become equal, the likelihood of two hydrophobic regions (which up to that point can be largely uncorrelated) becoming neighbors increases immensely and the role of the interface becomes less important [50].

7.3 Pre-Collapse of the Dendronized Polymers Below the Critical Temperature

As stated in the last section, the second spectral contribution originates at the critical temperature while only one species is observed below T_C. At 30 °C, the high field peak still possesses a Voigtian line shape (Fig. 7.1b). However, the peak is shifted to a lower magnetic field with respect to 5 °C, which is indicative of a slightly decreased hyperfine coupling constant.

In the following, the single spectral species below T_C is denoted C. The hyperfine coupling constants for this species a_C are depicted in Fig. 7.3a (closed symbols). For comparison, hyperfine coupling data for TEMPO in a purely aqueous environment a_{aq} are displayed (orange curve). These values decrease linearly with temperature in accordance with the decreasing dielectric constant of water.

At temperatures far below T_C, all hydrated polymers provide a slightly more hydrophobic environment than pure water since a_C falls $\sim$0.1 MHz below a_{aq}. This deviation could originate from the influence of the less hydrophilic polymer

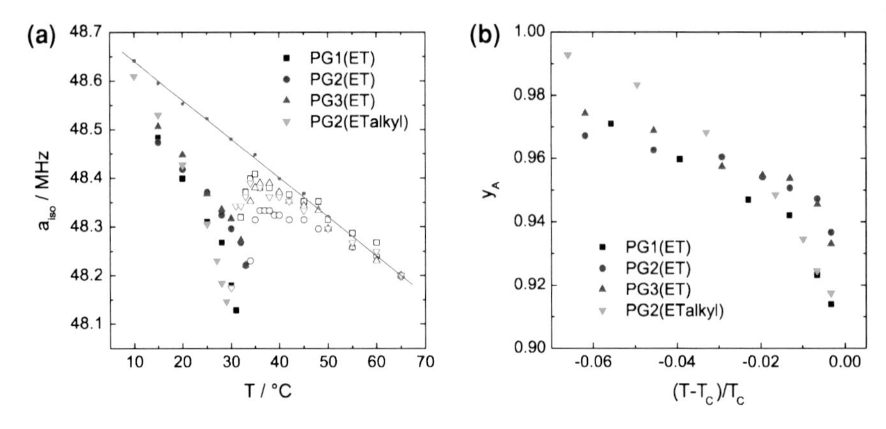

Fig. 7.3 **a** Hyperfine coupling constants of the single spectral component a_C (closed symbols, below T_C) and of the hydrophilic spectral component a_A (open symbols, above T_C) as a function of temperature. The hyperfine coupling constants of 0.2 mM TEMPO in pure water a_{aq} are depicted by orange squares and fitted with a straight line. **b** Variation of the fraction of TEMPO in a hydrophilic environment y_A below T_C as a function of the reduced temperature. The temperature dependent values are calculated by $y_A(T) = (a_C(T) - a_B(65\ °C))/(a_{aq}(T) - a_B(65\ °C))$. Reprinted with permission from [75]. Copyright 2011 WILEY-VCH Verlag GmbH & Co. KGaA, Weinheim

chains. Towards T_C, the deviation from the aqueous reference curve grows stronger with a maximum difference >0.3 MHz for PG1(ET) and PG2(ETalkyl).

The deviation from the purely aqueous solution is a measure for the ratio of hydrophilic and hydrophobic regions. In Fig. 7.3b, the fraction of spin probes in hydrophilic regions y_A is plotted vs the reduced temperature. A continuous decrease is observed with increasing temperature. The apparently increasing hydrophobicity originates from a partial dehydration of the polymers/dendrons as water becomes an increasingly poor solvent for the polymer. The spin probes exhibit a fast dynamic exchange between the small fraction of dehydrated regions (<9%) and the overwhelmingly large hydrophilic regions. This can be interpreted in terms of fast dynamic fluctuations of hydrated and collapsed polymer regimes. These fast dynamic fluctuations in the vicinity of the critical temperature were previously observed in pNiPAAm gels with dynamic light scattering [51, 52].

Up to a reduced temperature of -0.02, the decrease of y_A can be fitted with a straight line. Above this temperature, within 4 °C below T_C, the curve drops more steeply. In analogy to the EPR data above T_C, a second process governs the dehydration in this narrow temperature regime. In a second analogy, the dehydration of the polymer is governed by the hydrophobicity of the core. It is most efficient for PG2(ETalkyl) and PG1(ET), while PG2(ET) and PG3(ET) give rise to a less pronounced decrease of y_A.

In summary, an increase of the temperature below T_C leads to an increasing degree of dehydrated regions. This can be described as a partial collapse, which is not followed by an aggregation of several polymer chains or as an intrachain contraction preceding the interchain aggregation that takes place above T_C. This partial dehydration is accompanied by a decrease the hydrodynamic radius R_h of the polymer, which can be followed with light scattering. Wu et al. observed a decrease of ~40% for hydrophobized pNiPAAm derivatives [53]. In a recent light scattering study on the mesoglobule formation of PG2(ET), Bolisetty et al. noticed a decrease of the hydrodynamic radius from 18 to 16 nm [54].

Slightly above T_C, the hyperfine couplings of the hydrophilic spectral species a_A also deviate strongly from a_{aq} (Fig. 7.3a, open symbols). In this temperature regime, many predominantly hydrophilic regions (swollen dendrons, bulk water) are located in close vicinity to dehydrated polymer regions that lead to a small hydrophobic 'admixture' to the EPR signal. In the course of the microphase separation these regions are efficiently separated from the hydrophobic regions due to the large scale rearrangements in the complex macromolecular system. At high temperatures, spin probe species A is thus again located in a nanoscopic environment that resembles bulk water.

7.4 Formation and Influence of a Dense Peripheral Polymeric Layer

7.4.1 Results

In the previous Sects. 7.2 and 7.3, all aqueous polymer solutions were ramp heated to the maximum temperature (>10 K min^{-1}), and the EPR spectra were recorded upon a stepwise decrease of the temperature. In this section, these data are compared to samples that were heated with a considerably lower rate (<1 K min^{-1}). The EPR spectra were recorded during the heating process. Spectra measured upon cooling did not show a significant hysteresis in line with previous light scattering studies on PG2(ET) [54].

In Fig. 7.4, the hyperfine coupling values of the hydrophobic spectral component a_B are displayed for the slowly heated and ramp heated solutions. In essence, no dependence on the heating rate is observed for any dendronized polymer. One has to keep in mind that this spin probe species is in fast dynamic exchange between hydrophilic and hydrophobic regions. Hence, the apparent hyperfine coupling depends on the residence time of the spin probe in hydrophilic and hydrophobic regions. In that sense, it is a quantitative measure for the fraction of these regions itself. The observation of similar coupling constants implies that the ratio of hydrophilic and hydrophobic regions in the *immediate* environment of the spin probe is not affected by the annealing history. One has to stress that this observation is valid for the *nanoscopic* scale of several nm to understand the changes between fast and slow heating on different length scales.

In contrast, the fraction x_A of the spectral component S_A is severely affected by the rate of heating (Fig. 7.5a–c). For all polymers PG1–3(ET), the fraction of this

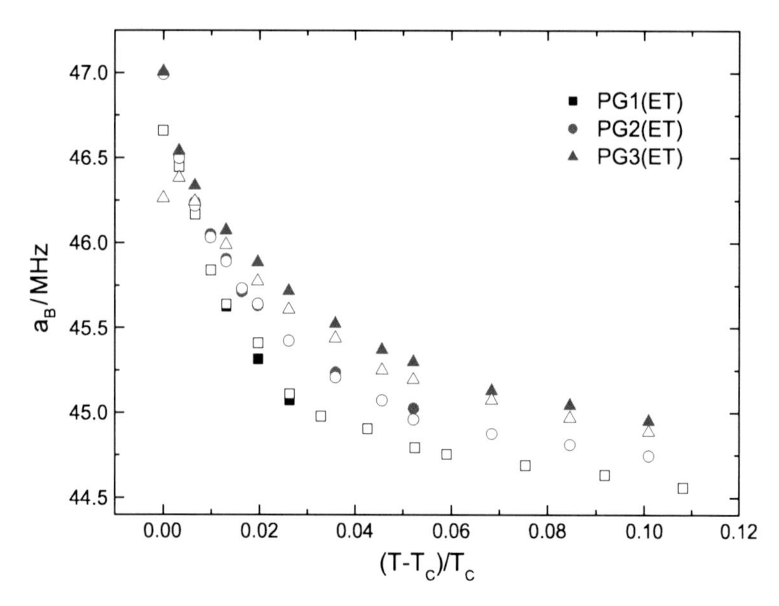

Fig. 7.4 Hyperfine coupling constants a_B above T_C as function of the reduced temperature for 0.2 mM TEMPO in 10 wt% aqueous solutions of PG1(ET), PG2(ET), and PG3(ET). Closed symbols represent data points obtained by ramp heating (>10 K min^{-1}), a slow heating rate (<1 K min^{-1}) gave rise to hyperfine couplings depicted by open symbols. Reprinted with permission from [76]. Copyright 2011 American Chemical Society

hydrophilic species was significantly increased at temperatures far above T_C, when the solutions were heated by a slow rate. This deviation is far more pronounced for the second and third generation dendronized polymers, while only small changes are observed for PG1(ET). The spectra of TEMPO in PG2(ETalkyl) are hardly affected by heating rate (data not shown).

The fraction x_A depends on the volumetric ratio of the collapsed/collapsing regions in the polymer and the hydrophilic bulk water and swollen polymer regions that are not in exchange with collapsed regions. This is a direct consequence of the amphiphilicity of the spin probe that chooses its environment depending on its hydrophobicity and its availability. Thus, an increase of the polymer concentration leads to a linear increase of the spectral species B (data not shown). Hence, an increased fraction x_A can be interpreted by an increased fraction of (purely) hydrophilic regions in the sample.

In combination with the observed trends for the hyperfine coupling values, it can be concluded that the heating rate affects the polymer solutions on the *microscopic* scale, while the *nanoscopic* environment of the spin probes remains unaffected. Irrespective of the heating rate, the obtained EPR spectra do not change with time. Both spin probe environments are stable for at least one hour as checked at a temperature of 40 °C.

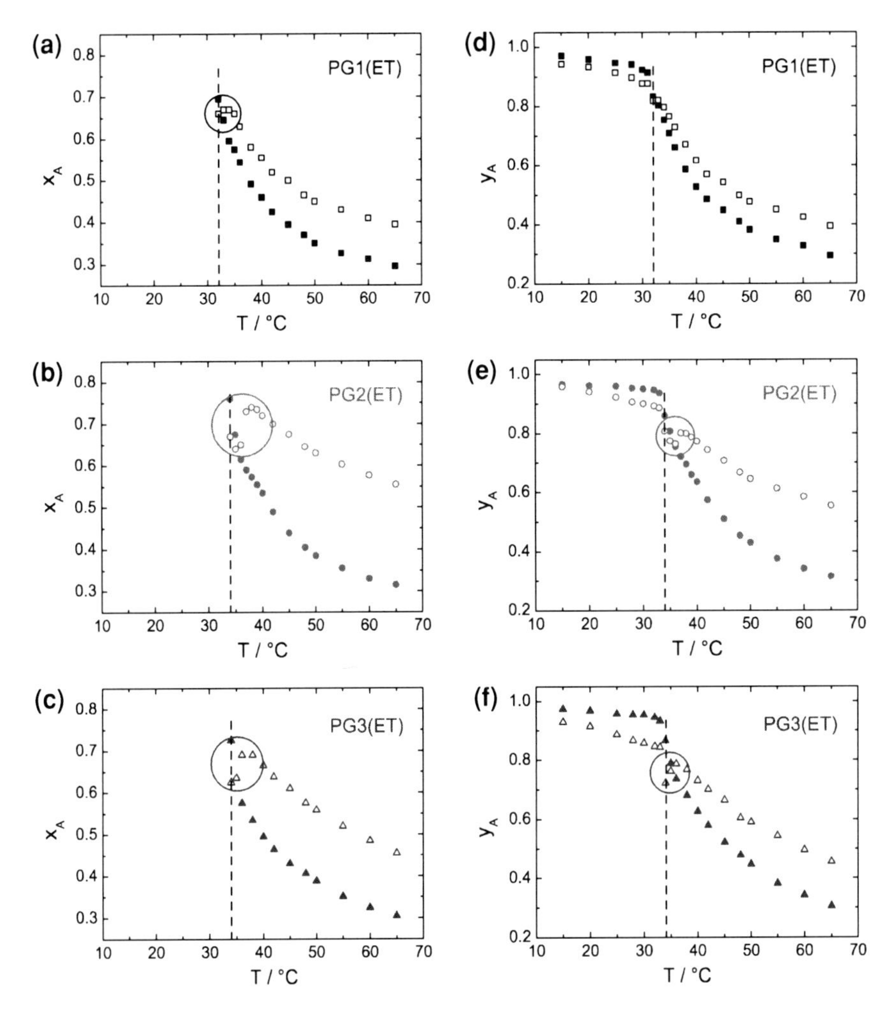

Fig. 7.5 **a–c** Fraction x_A of the hydrophilic spectral component S_A for 0.2 mM TEMPO in 10 wt% solution of PG1-3(ET) upon fast heating (closed symbols) and slow heating (open symbols). **d–f** Fraction of TEMPO in hydrophilic regions y_A as a function of temperature. The critical temperature T_C is marked by a dashed line. The local increase of S_A and y_A at temperatures slightly above T_C and at a slow heating rate is highlighted by circles. Reprinted with permission from [76]. Copyright 2011 American Chemical Society

After having described the general trend of the y_A curves, they are discussed in detail in this paragraph. At temperatures only slightly above T_C, a peculiar effect is observed upon slow heating, which is most pronounced for PG2(ET). Up to 35 °C (1 K above T_C), the ratios x_A are not affected by the heating rate. But while upon fast heating x_A decreases further as expected, a significant increase from 0.64 (35 °C) to a local maximum of 0.74 (38 °C) is observed in the case of slow heating. Only at higher temperatures, the expected decrease is observed.

This local maximum can be seen for all dendronized polymers PG1–3(ET). The magnitude of this feature is reflected in the deviation of x_A at high temperatures. At first sight, the mere occurrence of this feature is very surprising, since the hyperfine values suggest a steady formation of an increasing number of hydrophobic regions that are in exchange with hydrophilic regions. Here, it should be noted that, as mentioned above, the microscopic environment (sampled by x_A), is not reflected in the nanoscopic environment (sampled by a_B).

The overall fractions of (exchanging and non-exchanging) hydrophilic regions y_A bear all temperature-dependent features of x_A, since the hyperfine coupling is not affected by the heating rate (Fig. 7.5d–f). At T_C, the fractions y_A are not affected strongly by the heating rate. In fact, the fraction of hydrophilic regions is slightly decreased when the solution is heated with a slow rate. Further heating imposes a strong deviation in a narrow temperature interval of only 4 K, since a steep monotonous decrease is observed for a fast heating rate while the fraction of hydrophilic regions increases upon slow heating. At higher temperatures, both curves steadily decrease with the fast heating curve exhibiting a slightly steeper slope. At temperatures far above T_C, this leads to a dramatic difference. While at a fast heating rate 70% of all spin probes are located in collapsed, hydrophobic regions of PG2(ET), this value drops by one third to 45% when the solution is heated at a slow rate.

7.4.2 Discussion

All observed changes with respect to the rate of heating are related to the micro-/mesoscopic properties of the mesoglobules. These properties in the range of several hundred nanometers can be accessed by a variety of well-established methods. Up to date, the formation of mesoglobules in various thermoresponsive polymers was characterized by light scattering [9, 53, 55, 56], fluorescence spectroscopy [57], light and electron microscopy [56, 58], and differential scanning calorimetry [12, 16] on this length scale. In a recent publication, dynamic and static light scattering were utilized to monitor the temperature-dependent aggregation of PG2(ET) [54].

Mesoglobules are formed by association of individual polymer chains to multichain aggregates. At a certain, well-defined size, this association stops and nearly monodisperse, stable globular aggregates with sizes of 50 to several hundred nanometers are formed. Various explanations were proposed for the stability of a mesoglobule. In a first approach, the stability of the dispersion was attributed to electrostatic repulsion between particles, which could either be caused by charged polymer units or by associated salt ions [59, 60]. According to a common opinion, the more hydrophilic parts of the polymer are located preferably at the periphery of the mesoglobule in direct contact to the surrounding water and provide steric stabilization [61, 62]. This theory is supported by the fact that the copolymerization of the amphiphilic monomer with a small percentage of hydrophilic units leads to an increased stability of the mesoglobules.

Wu et al. however, observed that mesoglobules based on NiPAAm are also stabilized by hydrophobic comonomers [53]. According to their suggestion, the hydrophobic units promote intrachain contraction and harden the mesoglobules, thus slowing down chain motion. Hence, the interaction time between two colliding mesoglobules is not sufficient to induce a permanent chain entanglement [63]. Hence, they are protected from aggregation. This *viscoelastic* effect was already suggested previously to be one possible reason for the stability of homopolymer mesoglobules [56, 64].

It is well established that the size of the mesoglobules for a given polymer depends on three parameters: the polymer concentration, the temperature, and the rate of heating. For various amphiphilic polymers including PG2(ET) it was found that the size increases with increasing polymer concentration while it decreases at faster heating rates.

Though commonly observed, the dependence of the mesoglobule size on the heating rate is not studied in detail. Wu et al. proposed that on fast heating the intrachain contraction dominates the interchain aggregation [53]. The bulk of literature believes that a faster heating leads to a higher degree of vitrification of the particle's core [54, 55, 57], but only Van Durme et al. found conclusive experimental evidence for a partial vitrification of the polymer-rich phase [12]. However, modulated temperature DSC does not provide localized information about the vitrification. Along the same line, the hydrophobic parts of the polymers are commonly assumed to densify the inner part of the mesoglobule. Although theoretical studies provide support for this suggestion [61], there is no experimental evidence that the hydrophobic parts really self-assemble in the core of the sphere.

In the following, a different model is proposed, which provides an explanation for the obtained EPR data. Further, it is in agreement with the vast majority of published results on the formation of mesoglobules.

When the dendronized polymer is heated slowly to high temperatures, far more TEMPO molecules remain located in a hydrophilic environment than in the case of fast heating. Let us assume that a slow formation of the mesoglobule results in the expulsion of more spin probes into the aqueous phase. In this case, they would diffuse back into the mesoglobule since (thermodynamically) the ratio of the amphiphilic spin probes in hydrophilic and hydrophobic regions is only governed by the volumetric ratio of these regions. Thus, after a certain diffusion time, the ratio between spin probes in hydrophilic and hydrophobic regions would equilibrate again and show a heating-rate independent behavior. In contrast to this picture, the CW EPR spectra remain unaffected—irrespective of the heating rate—for at least one hour once the mesoglobules are formed.

Thus one has to propose a second effect that prohibits this probe diffusion into the hydrophobic regions of the mesoglobule. In the literature on responsive hydrogels, the so-called *skin barrier effect* is well known [1, 65, 66]. The collapse of such a gel results in the densification of its surface due to the expulsion of water.

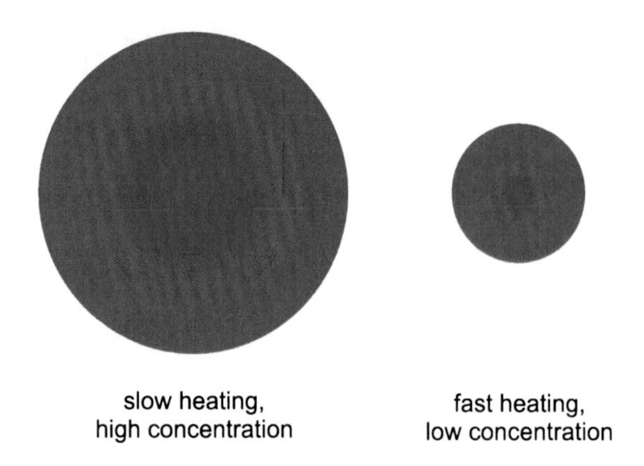

Scheme 7.2 Depiction of the skin effect in mesoglobules of different sizes. When increasing the temperature above a certain threshold, an impermeable outer polymeric layer is formed as the swollen polymer is dehydrated and densified (*orange*). For large polymer aggregates which are formed by slow heating and at high polymer concentration, a large amount of water is entrapped in the core (*blue*). A fast heating or a low polymer concentration result in smaller aggregates with a higher surface to volume ratio and the entrapment of less water

This collapsed surface becomes impermeable for water, which is still incorporated in the core of the gel. Thus, the hydrogel is kinetically trapped in a semi-collapsed state. The skin barrier effect could account for the observed effects regarding different heating rates in this study and in the mesoglobule formation in general. It also provides an explanation for effects that are related to the polymer concentration and the molecular weight of the formed aggregates as detailed in the next paragraphs.

Upon slow heating and at a higher polymer concentration, bigger mesoglobules are formed since interchain aggregation is promoted. For this case, let us first assume that the formation of an impermeable outer polymeric layer is not affected by the heating rate. Still, the skin barrier effect would be more pronounced in the bigger mesoglobule since the volume to surface ratio of a sphere increases linearly with its radius R. In other words, a larger mesoglobule possesses a larger core which is not able to collapse once a dense outer layer is formed. This is depicted schematically in Scheme 7.2.

In this framework, the increased fraction of spin probes in hydrophilic regions upon slow heating can be naturally explained. On the one hand, the diffusion of TEMPO from bulk water into the mesoglobule is prohibited due to the impermeable outer polymeric layer. This is true for both fast and slow heating, and should not affect the ratio of the spin probes. On the other hand, spin probes are trapped in the aggregates, which form a closed system. In case of large aggregates, this system contains a higher volume fraction of water, which leads to a higher fraction of spin probes in a hydrophilic environment. For small aggregates, the fraction of entrapped water is much lower and more spin probes are located in hydrophobic regions.

In this context, two issues need to be addressed. First, the hyperfine coupling values of the hydrophobic spectral compound a_B remain unaffected by the amount of entrapped water, though they provide a direct measure of the hydrophilic and hydrophobic fractions in the direct surrounding of the spin probe ($\sim$5 nm). This is a hint that a microphase separation takes place inside the mesoglobule. At the periphery, spin probes experience the environment of chains that are in the process of dehydration and are not affected by the entrapped water. In the core, the spin probes are placed in fully water swollen polymers or even bulk water in a range greater than their mean free pathway. Second, the fraction of spin probes in a hydrophobic environment grows larger when the temperature is increased. This could be explained by two effects. With increasing temperature there is a stronger tendency for the OEG chains to become more hydrophobic, which may lead to a larger volumetric ratio of dense polymeric regions ('skin thickness'). Further, every change of the temperature also leads to a reorganization of the polymer chains and dendrons in the mesoglobules and affects the surrounding of the spin probes. In the course of this reorganization, also small amounts of water may be released from the mesoglobule. This assumption is corroborated by light scattering studies that observed a shrinking of the mesoglobules above T_C [54, 55]. At a given temperature, however, the system remains in metastable state.

The 'micro' skin barrier effect proposed here explains, why the mesoglobules are stable in size though being in non-equilibrium conditions. At a given temperature and irrespective of the heating rate, the size of the mesoglobule remains constant, even if the polymer solution is diluted to substantially lower concentrations [55, 56]. It also accounts for the EPR spectra being time-independent at a certain temperature. The mesoglobules do not undergo structural changes and spin probe diffusion in and out of the aggregates is prohibited.

Under the assumption of a skin barrier and its increasing effect on larger mesoglobules, a large aggregate would approach the characteristics of a hollow (or rather water-filled) sphere. The collapsed chains give rise to a high polymer density at the periphery of the sphere, while the swollen polymers in the core exhibit a significantly decreased density. This problem can in principle be assessed by light scattering, since the ratio of the radius of gyration and the hydrodynamic radius R_g/R_h assumes characteristic values depending of the shape of the mesoglobule. For a uniform sphere, a value of 0.775 is expected, while an infinitely shallow hollow sphere gives rise to a characteristic ratio of $R_g/R_h = 1$ [67]. Thus, a mesoglobule with a denser outer part should exhibit an R_g/R_h between these two values.

Opposing the picture of a hollow sphere, Bolisetty et al. observed a ratio of $\sim$0.7 [54]. Kujawa et al. noted that values below 0.78 are obtained, characteristic of a molten globule with a denser core [57]. However, of all seven reported R_g/R_h ratios in their paper, only two values fall below a threshold of 0.77. Aseyev et al. stated that the ratio fluctuates around a value of 0.775, but under closer inspection, a clear linear increase of the ratio with increasing molecular weight of the formed aggregates is visible (Fig. 9 in Ref [55]). In fact, one can even find clear support for the proposed hollow sphere formation of large aggregates in this most detailed study with 20 different values for a variety of molecular weights. For small aggregates with low molecular weights $1 \cdot 10^8$ g mol^{-1}), the obtained values indeed fluctuate around the characteristic value of a uniform sphere, while values around 0.9 are found for large aggregates ($>1.5 \cdot 10^9$ g mol^{-1}).

In the previous paragraphs, it was explained and reasoned how the skin barrier effect could account for an entrapment of water and how this would lead to a higher spin probe fraction in a hydrophilic surrounding. The next paragraphs will focus on a peculiar detail of the EPR collapse curves that was neglected in the discussion so far: A slow heating rate results in a slight *increase* of the spin probes fraction in hydrophilic regions at temperatures slighty above the critical temperature (cf. Fig. 7.5).

Utilizing nonradiative energy transfer, Kujawa et al. observed that in a narrow temperature window of a few K in the vicinity of T_C, fluid pNiPAAm mesoglobules were formed that merged and grew [56]. Upon further heating, these mesoglobules underwent a conversion from fluid particles to rigid spheres that were unable to merge or undergo chain exchange. Further, they observed a broad distribution of relaxation times, which they attributed to heterogeneities in the mesoglobule. In their study of the mesoglobule formation of PG2(ET), Bolisetty et al. observed a transition from reaction limited colloidal aggregation to diffusion limited colloidal aggregation at 38 °C [54]. This is the exact temperature of the observed local maximum of the spin probe fraction in a hydrophilic environment y_A. All these observations support the proposed skin barrier effect and constitute a first hint that the dense polymer layer is formed in the same narrow temperature regime, in which the counterintuitive increase of y_A is observed.

Finally, by FTIR studies on pNiPAAm it was found that in this small temperature regime, loosely bound water molecules are expelled from the polymer chain into bulk water [15, 68–70]. With this last piece of information one can construct a detailed explanation of the mesoglobule formation in the framework of the skin barrier effect:

In a narrow regime above T_C ($\sim$4 K), a dense, impermeable polymeric layer is formed by the dehydration of peripheral dendrons/polymers. During this dehydration process, one part of the incorporated spin probes is released from the mesoglobule into the bulk water phase. Thus, the fraction y_A slightly rises. Once expelled from the mesoglobule, they are unable to re-enter due to the skin barrier. For the same reason, entrapped water is forced to stay inside. As the temperature increases, interactions between polymer and water become less favorable. Thus, the entrapped water is microphase separated from the polymer chains in the formed mesoglobule, which altogether leads to a transformation into a hollow sphere-like structure.

So far, a general explanation of the skin barrier formation and its effect on the mesoglobules for different heating rates and polymer concentrations has been given. Now, the question needs to be addressed, why there is a dependence of this effect on the dendron architectures. As described earlier, the heating rate hardly imposes any effect on the EPR spectra for PG2(ETalkyl). Only a slight change is observed for PG1(ET), while the spectra in PG2(ET) and PG3(ET) exhibit considerable deviations.

Similar to the dehydration and the precollapse, the skin barrier effect is governed by the hydrophilicity of the core. Due to their hydrophilic ethylenoxide core, PG2(ET) and PG3(ET) possess an intrinsic water reservoir that can be entrapped in the mesoglobule. From another point of view, they can be approximated as a mixture of hydrophobic (dendron periphery) and hydrophilic (dendron core) compounds. While the hydrophobic parts are responsible for the polymer aggregation and the formation of a skin barrier, the water-swollen hydrophilic compounds are located in the mesoglobule. PG1(ET) lacks a hydrophilic core, and PG2(ETalkyl) possesses a core, which exhibits a higher hydrophobicity than the periphery of the dendron. In line with the increased dehydration of single polymer chains below the critical aggregation temperature, PG1(ET) and PG2(ETalkyl) contain considerably less associated water when the impermeable polymer skin is formed. Hence, less water is entrapped and the skin effect leads to less pronounced changes of the mesoglobules and the incorporated spin probes.

7.5 Conclusions

In conclusion, the thermal transition of thermoresponsive dendronized polymers with different cores could be characterized on a molecular scale by CW EPR spectroscopy. Already ~4 K below the critical aggregation temperature, dynamic inhomogeneities are observed, which originate from a partial intra-polymer collapse on the nanometer scale. At T_C, these structural inhomogeneities are formed between different polymers and trigger the aggregation of the complete polymer sample into mesoglobules.

While macroscopic turbidity measurements suggest a sharp phase transition of the polymer, this study reveals that the dehydration of the polymer chains proceeds over a temperature interval of at least 30 °C. It cannot be described by a single de-swelling process that would be expected for a thermodynamic phase transition. Rather, the dehydration should be viewed as a molecularly controlled non-equilibrium state and takes place in two steps. The local heterogeneities grow in size and polymer chain fluctuations slow down. Within ~7 K above T_C, the majority of the dehydration is completed and percolation for the fraction and volume of hydrophobic regions is reached. Heating the samples even higher only leads to an additional loss of residual water from the collapsed system.

Heating rate dependent changes of spin probe fractions in hydrophilic and hydrophobic environments were interpreted by the formation of a dense polymeric layer at the periphery of the mesoglobule, which is also formed in a narrow temperature range of ~4 K above T_C and prohibits the release of molecules that are incorporated in the polymer aggregate. Thus, a considerable amount of water is

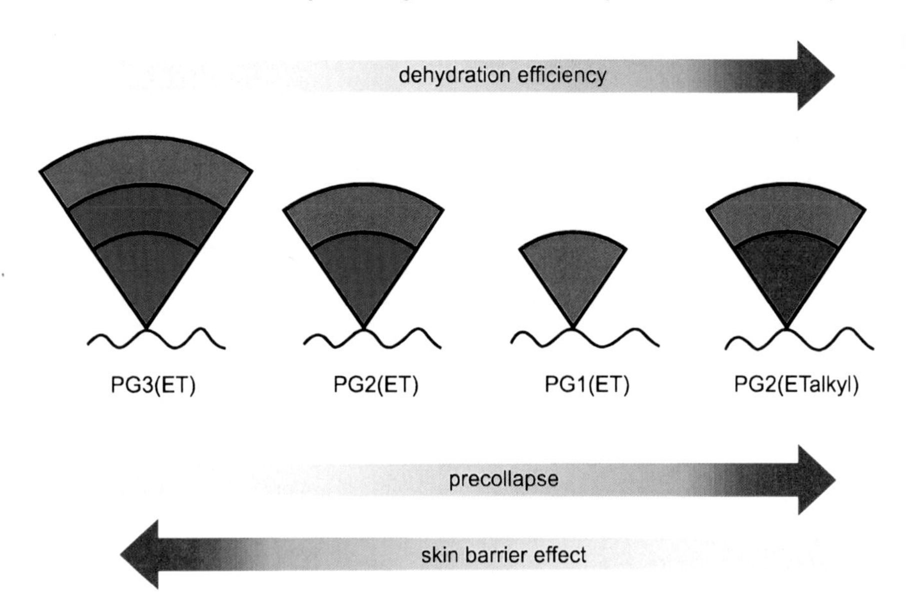

Scheme 7.3 Effect of the dendron architecture on the thermoresponsive properties of the polymers. A hydrophilic core (*blue* segments for PG2(ET) and PG3(ET)) decreases the dehydration efficiency of the mesoglobule and counteracts the formation of inhomogeneities below the critical temperature (pre-collapse). Large amounts of incorporated water give rise to a pronounced skin barrier effect that imposes significant changes with respect to the rate of heating and the polymer concentration. These effects are reverted by a hydrophobic core (red segment in PG2(ETalkyl)), which increases both the dehydration efficiency and the pre-collapse and diminishes the amount of entrapped water in the mesoglobule

entrapped in large aggregates, which are formed at low heating rates and at high polymer concentrations.

While the aggregation temperature mainly depends on the periphery of the dendrons, all other observed processes and effects are sensitive to the hydrophilicity of their inner parts. A hydrophobic core provides for a high dehydration efficiency of the formed mesoglobules. The dependence of these three processes on the molecular architecture is summarized in Scheme 7.3. Even below the aggregation temperature, it triggers a partial collapse of single polymer chains as water becomes an increasingly poor solvent. Further, the hydrophilicity of the core governs the amount of water that is incorporated in the mesoglobule and entrapped due to the impermeable skin barrier.

In summary, CW EPR is a particularly simple technique that provides unique information on the thermal transitions of amphiphilic polymers both on the molecular and on the mesoscopic level. Thus, it is an indispensible addition to the commonly applied characterization techniques such as light scattering. While light scattering is restricted to low polymer concentrations, high concentrations are favorable for EPR studies since the volume fraction of the polymer governs the distribution of the amphiphilic spin probe.

7.6 Materials and Methods

Materials. All thermoresponsive dendronized polymers were prepared by Wen Li. The synthetic procedures are reported elsewhere [24–26, 71]. 2,2,6,6-Tetramethylpiperidine-1-oxyl (TEMPO, 98%, Aldrich) was used as received. Distilled water was further purified by a MilliQ System (Millipore) to achieve a resistivity of 18.2 MΩ cm.

Sample Preparation. A 0.2 mM aqueous TEMPO solution was added to the appropriate amount of polymer to achieve a 10 wt% polymer solution. The high polymer concentration was chosen since the ratio of the amphiphilic spin probe in hydrophilic and hydrophobic environments strongly depends on the volume fraction of the collapsed polymer in water. No fundamental differences to diluted polymer solutions are expected, since turbidity measurements on comparable dendronized polymer systems showed that the aggregation temperature between highly diluted and concentrated solutions differs by only 2 K [24].

EPR studies. All EPR spectra were recorded by Wen Li on a Miniscope MS200 (Magnettech, Berlin, Germany) benchtop spectrometer working at X-band (~9.4 GHz) with a modulation amplitude of 0.04 mT, a sweep width of 8 mT, a sweep time of 30 s, and a microwave power of 12.5 mW. The temperature was adjusted with the temperature control unit TC H02 (Magnettech), providing for an electronic adjustment in steps of 0.1 K in the range of −170 to 250 °C. For the reported data in the first two parts of this chapter (Sects. 7.2 and 7.3), the polymer solutions were ramp heated (>30 K min^{-1}) to a maximum temperature of 65 °C, then decreased stepwise for the actual measurements. In the second part, EPR spectra are shown, which result from a slow heating (<1 K min^{-1}) of the samples (Sect. 7.4).

Data analysis and interpretation. CW EPR spectral simulations were performed with a Matlab program utilizing the fast-motion routine of the Easyspin software package for EPR [72]. Above T_C, two spectral components A and B could be distinguished that add to the combined simulation S_{tot} in the appropriate fraction x_i ($S_{tot} = x_A S_A + (1 - x_A) S_B$). The isotropic hyperfine and g-values, a_{iso} and g_{iso}, are inherently coupled and affect each other. Thus, it is difficult to infer the precise g-value in two-component spectra by spectral simulations. In the regime relevant for species A, the relationship between a_A and g_A was determined by analysis of the single component nitroxide spectrum of 0.2 mM TEMPO in water at various temperatures, yielding a set of a_{aq} and g_{aq} pairs. The relation between a_B and g_B was then approximated by a linear extrapolation of the dataset. The overall fraction of spin probes in a hydrophilic environment y_A for temperatures above T_C was determined by

$$y_A(T) = x_A \frac{a_A(T) - a_B(65\ ^\circ\mathrm{C})}{a_{aq}(T) - a_B(65\ ^\circ\mathrm{C})} + (1 - x_A) \frac{a_B(T) - a_B(65\ ^\circ\mathrm{C})}{a_{aq}(T) - a_B(65\ ^\circ\mathrm{C})}, \tag{7.1}$$

assuming that a_B (65 °C) represents the static hyperfine coupling limit of spin probes located in hydrophobic, collapsed regions. Below T_C, simulations were performed assuming a single spin probe species with parameters a_C and g_C. The overall fraction of spin probes in a hydrophilic environment y_A was calculated by

$$y_A(T) = \frac{a_C(T) - a_B(65\ ^\circ\mathrm{C})}{a_{aq}(T) - a_B(65\ ^\circ\mathrm{C})}. \tag{7.2}$$

The simulation procedure is detailed in Appendix A.4.1 on the example of PG1(ET).

References

1. Schild HG (1992) Prog Polym Sci 17:163–249
2. de las Heras Alarcón C, Pennadam S, Alexander C (2005) Chem Soc Rev 34:276–285
3. Yerushalmi R, Scherz A, van der Boom ME, Kraatz H-B (2005) J Mater Chem 15:4480–4487
4. Jia Z, Chen H, Zhu X, Yan D (2006) J Am Chem Soc 128:8144–8145
5. Kumar A, Srivastava A, Galaev IY, Mattiasson B (2007) Prog Polym Sci 32:1205–1237
6. Chen H, Jia Z, Yan D, Zhu X (2007) Macromol Chem Phys 208:1637–1645
7. Helms B, Fréchet JMJ (2006) Adv Synth Catal 348:1125–1148
8. Wu C, Zhou SQ (1995) Macromolecules 28:5388–5390
9. Wu C, Zhou SQ (1995) Macromolecules 28:8381–8387
10. Wu C, Zhou SQ (1996) Phys Rev Lett 77:3053–3055
11. Wang X, Qiu X, Wu C (1998) Macromolecules 31:2972–2976
12. Van Durme K, Verbrugghe S, Du Prez FE, Van Mele B (2004) Macromolecules 37:1054–1061
13. Luo S, Xu J, Zhu Z, Wu C, Liu S (2006) J Phys Chem B 110:9132–9139
14. Ono Y, Shikata T (2006) J Am Chem Soc 128:10030–10031
15. Cheng H, Shen L, Wu C (2006) Macromolecules 39:2325–2329
16. Van Durme K, Van Assche G, Aseyev V, Raula J, Tenhu H, Van Mele B (2007) Macromolecules 40:3765–3772
17. Ono Y, Shikata T (2007) J Phys Chem B 111:1511–1513
18. Keerl M, Smirnovas V, Winter R, Richtering W (2008) Angew Chem Int Ed 47:338–341
19. Schlüter AD, Rabe JP (2000) Angew Chem Int Ed 39:864–883
20. Zhang A, Shu L, Bo Z, Schlüter AD (2003) Macromol Chem Phys 204:328–339
21. Schlüter AD (2005) Top Curr Chem 245:151–191
22. Frauenrath H (2005) Prog Polym Sci 30:325–384
23. Rosen BM, Wilson CJ, Wilson DA, Peterca M, Imam MR, Percec V (2009) Chem Rev 109:6275–6540
24. Li W, Zhang A, Feldman K, Walde P, Schlüter AD (2008) Macromolecules 41:3659–3667
25. Li W, Zhang A, Schlüter AD (2008) Chem Commun 5523–5525
26. Li W, Wu D, Schlüter AD, Zhang A (2009) J Polym. Sci, Part A: Polym Chem 47:6630–6640
27. Dormidontova EE (2004) Macromolecules 37:7747–7761
28. Ober CK, Cheng SZD, Hammond PT, Muthukumar M, Reichmanis E, Wooley KL, Lodge TP (2009) Macromolecules 42:465–471
29. Cady F, Qian H (2009) Phys Biol 6:036011
30. Brutlag D, Kornberg A (1972) J Biol Chem 247:241–248
31. Drobny GP, Long JR, Karlsson T, Shaw W, Popham J, Oyler N, Bower P, Stringer J, Gregory D, Mehta M, Stayton PS (2003) Annu Rev Phys Chem 54:531–571

32. Schmidt-Rohr K, Spiess HW (1994) Multidimensional solid-state NMR and polymers. Academic Press, London
33. Saalwächter K (2007) Prog Nucl Magn Reson Spectrosc 51:1–35
34. Schlick S (ed) (2006) Advanced ESR methods in polymer rsearch. Wiley-Interscience, Hoboken
35. Schmidt-Rohr K, Spiess HW (1991) Phys Rev Lett 66:3020–3023
36. Heuer A, Wilhelm M, Zimmermann H, Spiess HW (1995) Phys Rev Lett 75:2851–2854
37. Tracht U, Wilhelm M, Heuer A, Feng H, Schmidt-Rohr K, Spiess HW (1998) Phys Rev Lett 81:2727–2730
38. Hansen MR, Schnitzler T, Pisula W, Graf R, Müllen K, Spiess HW (2009) Angew Chem Int Ed 48:4621–4624
39. Ruthstein S, Raitsimring AM, Bitton R, Frydman V, Godt A, Goldfarb D (2009) Phys Chem Chem Phys 11:148–160
40. Hinderberger D, Spiess HW, Jeschke G (2010) Appl Magn Reson 37:657–683
41. Dockter C, Volkov A, Bauer C, Polyhach Y, Joly-Lopez Z, Jeschke G, Paulsen H (2009) Proc Natl Acad Sci USA 106:18485–18490
42. Hinderberger D, Schmelz O, Rehahn M, Jeschke G (2004) Angew Chem Int Ed 43:4616–4621
43. Owenius R, Engstrom M, Lindgren M, Huber M (2001) J Phys Chem A 105:10967–10977
44. Junk MJN, Jonas U, Hinderberger D (2008) Small 4:1485–1493
45. Knauer BR, Napier JJ (1976) J Am Chem Soc 98:4395–4400
46. Kovarskii AL, Wasserman AM, Buchachenko AL (1972) J Magn Reson 7:225–237
47. Atherton NM (1993) Principles of electron spin resonance. Ellis Horwood, New York
48. Derkaoui N, Said S, Grohens Y, Olier R, Privat M (2007) J Colloid Interface Sci 305:330–338
49. Chaikin PM, Lubensky TC (1995) Principles of condensed matter physics. Cambridge University Press, Cambridge
50. Note that at the chosen polymer and spin probe concentrations $\sim$30% of the spin probes reside in hydrophilic regions even at high temperatures. Therefore, the potential percolation point at $y_A \sim 0.5$ appears when ($0.5/0.7 \approx 0.7$) 70% of all probes taking part in the transition are trapped in hydrophobic regions.
51. Shibayama M, Takata S, Norisuye T (1998) Phys A 249:245–252
52. Shibayama M (1998) Macromol Chem Phys 199:1–30
53. Wu C, Li W, Zhu XX (2004) Macromolecules 37:4989–4992
54. Bolisetty S, Schneider C, Polzer F, Ballauff M, Li W, Zhang A, Schlüter AD (2009) Macromolecules 42:7122–7128
55. Aseyev V, Hietala S, Laukkanen A, Nuopponen M, Confortini O, Du Prez FE, Tenhu H (2005) Polymer 46:7118–7131
56. Gorelov AV, Du Chesne A, Dawson KA (1997) Physica A 240:443–452
57. Kujawa P, Aseyev V, Tenhu H, Winnik FM (2006) Macromolecules 39:7686–7693
58. Li W, Zhang A, Chen Y, Feldman K, Wu H, Schlüter AD (2008) Chem Commun 5948–5950
59. Chan K, Pelton R, Zhang J (1999) Langmuir 15:4018–4020
60. Balu C, Delsanti M, Guenoun P, Monti F, Cloitre M (2007) Langmuir 23:2404–2407
61. Vasilevskaya VV, Khalatur PG, Khokhlov AR (2003) Macromolecules 36:10103–10111
62. Baulin VA, Zhulina EB, Halperin A (2003) J Chem Phys 119:10977–10988
63. Picarra S, Martinho JMG (2001) Macromolecules 34:53–58
64. Dawson KA, Gorelov AV, Timoshenko EG, Kuznetsov YA, Du Chesne A (1997) Physica A 244:68–80
65. Matsuo ES, Tanaka T (1988) J Chem Phys 89:1695–1703
66. Yu H, Grainger DW (1993) J Appl Polym Sci 49:1553–1563
67. Burchard W (1983) Adv Polym Sci 48:1–124
68. Maeda Y, Higuchi T, Ikeda I (2000) Langmuir 16:7503–7509
69. Maeda Y, Higuchi T, Ikeda I (2001) Langmuir 17:7535–7539
70. Maeda Y, Nakamura T, Ikeda I (2001) Macromolecules 34:1391–1399

71. Li W (2010) Novel dendritic macromolecules with water-soluble. Thermoresponsive and amphiphilic properties, Doctoral Dissertation, ETH Zürich
72. Stoll S, Schweiger A (2006) J Magn Reson 178:42–55
73. Junk MJN, Li W, Schlüter AD, Wegner G, Spiess HW, Zhang A, Hinderberger D (2010) Angew Chem 122:5818–5823
74. Junk MJN, Li W, Schlüter AD, Wegner G, Spiess HW, Zhang A, Hinderberger D (2010) Angew Chem Int Ed 49:5683–5687
75. Junk MJN, Li W, Schlüter AD, Wegner G, Spiess HW, Zhang A, Hinderberger D (2011) Macromol Chem Phys 212:1229–1235
76. Junk MJN, Li W, Schlüter AD, Wegner G, Spiess HW, Zhang A, Hinderberger D (2011) J Am Chem Soc 133:10832–10838

Chapter 8
Conclusion

In this thesis, the functional structure of different amphiphilic macromolecules and oligomers was studied by CW and pulse EPR spectroscopy. Due to their structural heterogeneities on the molecular and nanoscopic level they can host and transport small molecules. By employing EPR-active probe molecules and ions, the *local* structure of these materials was examined from the unique perspective of guest molecules that are incorporated in the systems and directly experience all structural features relevant for the transport characteristics.

Human serum albumin exhibits the best-defined structure of all studied materials, which is still highly adaptable to fit the requirements of a myriad of ligands. The binding of fatty acids studied here originates from a sophisticated interplay between hydrophobic channels and charged anchor groups. Using spin-labeled fatty acids, structural information could be accessed that is directly related to this highly relevant transport function. It was found that the charged protein residues, which function as anchors for fatty acids, are distributed asymmetrically in the protein as suggested by the crystal structure (Chap. 3). In contrast, a remarkably symmetric entry point distribution of the fatty acid binding channels was found in contrast to the crystal structure, which suggests an asymmetric distribution for both, anchor and entry points. It can be concluded that the structure of these highly dynamic surface-exposed regions is not fully reflected by the crystal structure, which rather represents one entrapped conformer of the protein. A symmetric distribution of the binding sites' entry points facilitates the uptake and release of fatty acids and aids the optimized function of the protein as a transporter of fatty acids.

Star-shaped cholic acid oligomers also show a complex molecular self-organization due to the facial amphiphilicity of their building blocks. They can host and transport hydrophobic molecules in hydrophilic solutes and vice versa. Besides the inclusion of guest molecules, a preordered arrangement of incorporated 1,2,3-triazole groups induces a strong, chelate-like complexation of metal ions and leads to a close contact between metal and guest (Chap. 4). A ternary self-assembled

M. J. N. Junk, *Assessing the Functional Structure of Molecular Transporters by EPR Spectroscopy*, Springer Theses, DOI: 10.1007/978-3-642-25135-1_8,

system of oligomer, guest molecule, and metal ion bears a remarkable structural analogy to an active site in metal-containing enzymes. However, due to their limited size, the cholic acid derivatives cannot achieve a structural complexity that determines the pronounced binding selectivity and versatility of HSA. Further, the self-assembly of the metal ions is largely affected by slight structural variations of the molecule. Without structural restrictions, each triazole unit acts as a single entity, thus significantly reducing the coordination characteristics of the oligomer. The loose binding of the metal ions results in a loss of the metal–guest molecule contact and of the enzyme-like function.

Despite the absence of structural regularities as in biological systems, synthetic polymeric amphiphiles were found to exhibit surprisingly complex structures and functions. During the thermally induced phase separation, **hydrogels based on *N*-isopropylacrylamide** phase segregate into hydrophilic, water-swollen and hydrophobic, collapsed domains on the nanometer length scale. In contrast to sharp macroscopic volumetric changes, the phase separation proceeds over a significantly broader temperature range on this local scale (Chap. 5).

This broad transition range is also reflected in the efficiency of the NMR signal enhancement by **dynamic nuclear polarization (DNP)**. A gradual rather than a sharp decrease was found when the critical temperature of the spin-labeled hydrogel is approached, which is indicative for a slowly increasing fraction of dehydrated regions. In general, it was demonstrated that spin-labeled thermoresponsive hydrogels are suitable hyperpolarizing agents that allow for an efficient radical–solute separation (Chap. 6). The partial collapse due to microwave-induced heating can be prevented by DNP measurements at low temperatures. Hence, the full strength of these DNP agents is to be revealed.

Remarkably, the nanoscopic structural heterogeneities of the **thermoresponsive hydrogels** manifest themselves in an inhomogeneous catalytic activity. While the chemical decay of spin probes is significantly accelerated in hydrophilic regions, spin probes in hydrophobic cavities are protected from any decay reaction (Chap. 5). This pronounced change of functionality in a range of only few nanometers bears surprising analogies to enzymes, which also contain both catalytic reaction centers and protective sites. It is remarkable that a statistical ternary thermoresponsive copolymer exhibits a catalytic activity and specificity comparable to functional biomacromolecules that are optimized for a specific type of reaction.

In an analogous study, the nanoscopic structure of **thermoresponsive dendronized polymers** with different dendritic cores was characterized. Again, the phase separation results in local structural heterogeneities of hydrated and dehydrated regions. However, the observed heterogeneities of dendronized polymers are highly dynamic in contrast to the heterogeneities observed in the studied hydrogels, which are static on a time scale of at least two hours. This nicely illustrates the dependence of the polymer architecture on the local behavior. For hydrogels, the phase separation and encapsulation of the spin probes is governed by inhomogeneous chemical crosslinks. In dendronized polymers, physical crosslinks formed by extended side chains are responsible for these processes (Chap. 7).

Already $\sim$4 K below the critical aggregation temperature, a partial intra-polymer collapse leads to structural heterogeneities, which trigger the aggregation of the complete polymer sample into mesoglobules at the critical temperature. In analogy to the broad phase transition of the hydrogels, the dehydration of the polymer chains proceeds stepwise over a temperature interval of at least 30 °C and is facilitated by a hydrophobic dendritic core. The apparent hyperfine coupling and the relative fraction of spin probes in hydrophilic and hydrophobic regions allow for a direct comparison of the structure on different length scales. Structural differences originate from the kinetic entrapment of a non-equilibrium state, which results from the formation of an impermeable dense polymeric layer at the periphery of the mesoglobule (Chap. 7).

In summary, the local structure and function of a broad range of amphiphilic systems was characterized by EPR spectroscopy. The selective and sensitive paramagnetic spin probes revealed important and surprising features of the materials that are not or barely accessible by other techniques. The versatility of EPR spectroscopic methods allows to study a structure forming feature as general as amphiphilicity in all its different shades that can be found in biological and synthetic systems. It may thus be extended to many other systems of varying degrees of order and structure.

Appendix

A.1 The Functional Structure of Human Serum Albumin

A.1.1 Reduction of Spin-Labeled Fatty Acids to EPR-Inactive Hydroxylamines

Reduction with Phenylhydrazine

The doxyl groups of the spin-labeled fatty acids were reduced to hydroxylamines with phenylhydrazine following the reaction scheme of Lee and Keana (Scheme A.1) [1]. Phenylhydrazine (97%, Aldrich) was used as received. A 20 mM solution of phenylhydrazine in 0.1 M KOH was freshly prepared prior to each reduction (no coloration). 100 μl of 8.33 mM DSA (1 eq) in 0.1 M KOH were reacted with 25 μl of this solution (0.6 eq phenylhydrazine) at room temperature under argon atmosphere. The conversion of the reaction was followed by CW EPR spectroscopy (Fig. A.1). Optimum conversions were achieved after 30 min (16-DSA) and 45 min (5-DSA), and using a slight excess of the reducing agent (0.1 eq). Under these conditions, 93 ± 3% of 16-DSA and 5-DSA were converted into EPR-silent species. The resulting solution of 6.67 mM rDSA was used without purification. The final HSA–fatty acid samples were kept under argon to prevent re-oxidation of the hydroxylamine by oxygen.

The above-mentioned method provides the best reduction efficiency and was used for the preparation of all samples described in Chap. 3. However, phenylhydrazine is mutagenic, and remaining traces of this reagent could potentially alter the structure of HSA. Thus, a second reduction technique was chosen to confirm the obtained results and to ensure that the protein structure is not altered.

M. J. N. Junk, *Assessing the Functional Structure of Molecular Transporters by EPR Spectroscopy*, Springer Theses, DOI: 10.1007/978-3-642-25135-1,

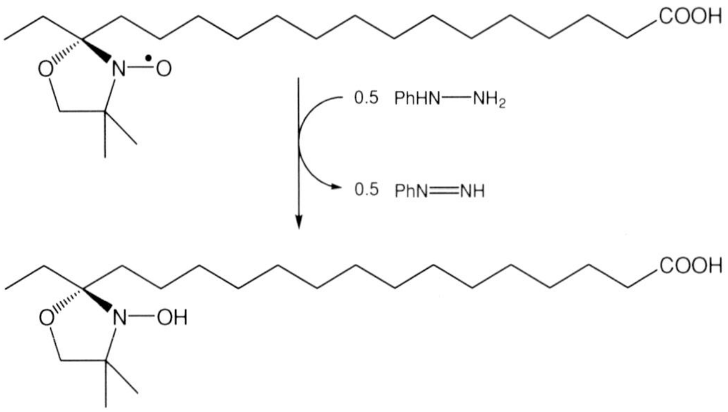

Scheme A.1 Reduction of 16-Doxylstearic Acid. Reprinted with permission from [12, 13]. Copyright 2010 WILEY-VCH Verlag GmbH & Co. KGaA, Weinheim

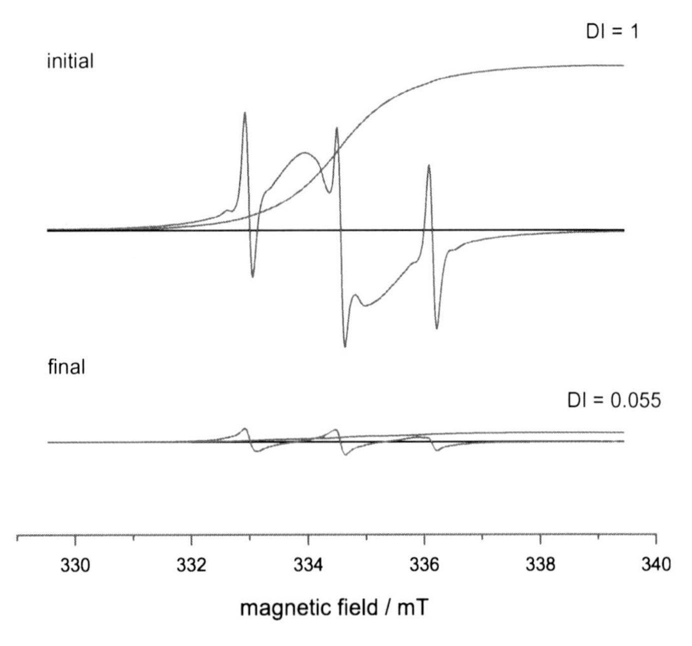

Fig. A.1 CW EPR spectra and double integrals of 16-DSA before (*red*) and after (*blue*) reduction with phenylhydrazine. The double integrals (DI) of the spectra correspond to the radical concentration and can be compared directly as the spectra were recorded under the same conditions. This allows determining the conversion (93%), taking into account the altered fatty acid concentrations due to addition of the phenylhydrazine solution. Reprinted with permission from [12]. Copyright 2010 WILEY-VCH Verlag GmbH & Co. KGaA, Weinheim

Reduction with Ascorbic Acid

The synthetic scheme of Paleos and Dais was followed [2]. Ascorbic acid (99%, Aldrich) was used as received. 2.68 mg (7.0 mmol) 16-DSA was reacted with 12.33 mg (70 mmol) ascorbic acid in 10 mL 0.1 M KOH at room temperature under argon atmosphere. Though the reaction was described stoichiometrically quantitative in the original publication, only a 20-fold excess of the reducing agent provided for an almost quantitative conversion to the hydroxylamine after 30 min, as checked by CW EPR. The mixture was neutralized with HCl (conc.) and extracted with diethylether (3 $\times$ 10 mL). The organic phase was extracted with deionized water (10 mL) and dried with $MgSO_4$. The solvent was removed by evaporation. The yield was 60%. During the purification process, 15 $\pm$ 5% of the fatty acids were reoxidized to EPR-active nitroxides.

After preparation of a 6.67 mM solution of 16-rDSA in 0.1 M KOH, HSA–fatty acid complexes were prepared by mixing with 16-DSA and HSA as described in the main text. Both CW EPR and DEER data (not shown) are in agreement with the results obtained with the phenylhydrazine reduced fatty acid. Hence, potential phenylhydrazine traces do not induce changes of the protein structure that can be detected by EPR.

Stearic Acid as EPR-Inactive Molecule
Stearic acid as a diamagnetic molecule can be also used as a spin diluting additive. In advantage to rDSA, it allows a direct comparison with the crystal structure. However, it exhibits slight structural deviations to DSA concerning both steric and polar effects. This could lead to different binding preferences and thus to a non-statistic distribution of the DSA molecules among the fatty acid binding sites.

Another problem relates to its low solubility in water. In contrast to DSA and rDSA, it is not soluble in 0.1 M KOH at a concentration of 6.67 mM. A relatively homogeneous suspension is obtained, which can be admixed to DSA and HSA. Up to a ratio of 1:2:2 (HSA:SA:DSA), all stearic acid molecules are complexed by HSA as indicated by the CW EPR spectra and the clear solution obtained. Above a ratio of 1:2:2, the CW EPR spectra do not change and a precipitate is observed, indicating that no further stearic acid molecules are absorbed.

The obtained DEER data are in agreement with samples containing reduced DSA (data not shown). This could indicate that the structural deviations do not lead to different preferential binding sites. Given the fact that the DEER data do not undergo substantial changes at any condition, a second possible explanation implies that the fatty acids are distributed very homogeneously in the protein and binding site preferences do not manifest in different distance distributions. This is elaborated in Sect. 3.2.

A.1.2 CW EPR Studies With Paramagnetic DSA Without Spin Dilution

For comparison, CW EPR spectra of HSA–16-DSA complexes without diamagnetic fatty acids were recorded at different ratios (Fig. A.2). Different binding affinities and binding site preferences of the diamagnetic and the paramagnetic DSA molecules would manifest in different ratios of bound and unbound species compared to the spin-diluted CW EPR spectra (Fig. 3.2a).

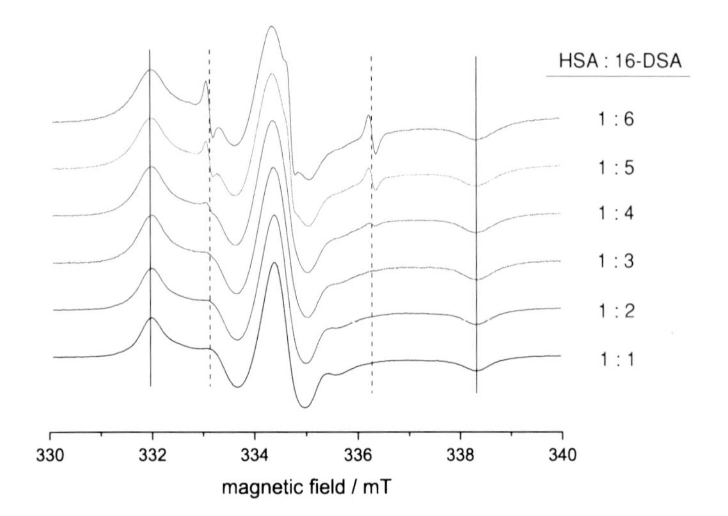

Fig. A.2 CW EPR spectra of HSA and 16-DSA at different protein–fatty acid ratios ($T = 298$ K). The characteristic signatures of DSA bound to HSA are marked by *solid lines*. The signatures of free fatty acids in solution are marked by *dashed lines*. Reprinted with permission from [12, 13]. Copyright 2010 WILEY-VCH Verlag GmbH & Co. KGaA, Weinheim

Large differences can be excluded, since the series of spectra exhibit only small differences. For the samples containing solely 16-DSA, up to three fatty acids are complexed without the signature of an unbound species. At this ratio, only a negligible amount of fatty acid is not taken up, when a mixture of 16-rDSA and 16-DSA is applied (cp. ratio 1:1:2, Fig. 3.2a). A slightly more prominent signal of the unbound fatty acid at higher HSA–DSA ratios originates from the fact that the signal of the bound species is slightly broadened due to dipolar couplings and Heisenberg spin exchange and is thus decreased in its height. The ratios of the double integrals (which correspond to the total concentrations) of both species are in agreement, with deviations within the experimental error margin at all ratios.

A.1.3 CW EPR Studies of Fatty Acid Free HSA

Under physiological conditions, between 0.1 and 2 fatty acid molecules are bound to HSA. When studying the complexation of fatty acids by HSA, these natural intruders could obscure or distort the results. For this reason, HSA batches without residual fatty acids were used in many studies dealing with this topic.

In Fig. A.3, CW EPR spectra of HSA–16-DSA mixtures at different ratios are shown. For this reason, fatty acid free (f.a.f.) HSA (Sigma, <0.007% FA, purification by agarose gel electrophoresis) was used as received.

Surprisingly, a small signature of free fatty acid is observed even at a ratio of 1:2. In general, the spectra with f.a.f. HSA correspond to spectra of non-denaturated HSA complexed with two additional fatty acids (Fig. A.2).

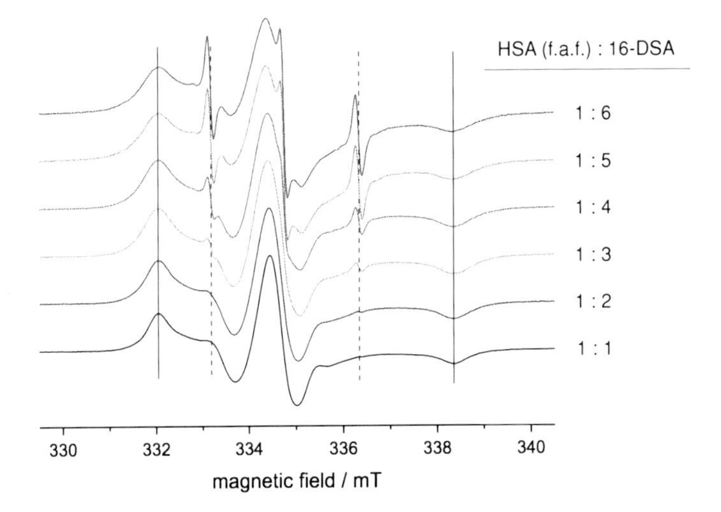

Fig. A.3 CW EPR spectra of fatty acid free HSA and 16-DSA at different protein–fatty acid ratios ($T = 298$ K). The characteristic signatures of DSA bound to HSA are marked by *solid lines*. The signatures of free fatty acids in solution are marked by *dashed lines*. Reprinted with permission from [12, 13]. Copyright 2010 WILEY-VCH Verlag GmbH & Co. KGaA, Weinheim

Though essentially fatty acid free albumin should be capable to bind more or at least equal amounts of fatty acids, a decreased fatty acid binding is observed compared to non-denaturated HSA. This could be due to a partial degeneration of the protein caused by the purification process, which could result in the loss of one (or several) fatty acid binding sites.

For this reason, non-denaturated HSA was used throughout this study. Potentially residual fatty acid molecules were regarded as a far less severe problem, since they can exchange with EPR-active fatty acids and do not prohibit the desired uniform distribution of DSA among all possible binding sites. This fast exchange of even tightly bound fatty acids is affirmed by a ready exchange of DSA and rDSA. Comparable CW EPR and DEER spectra were obtained irrespective of simultaneous or consecutive addition to the protein (data not shown).

A.1.4 No Dimers in Solution Observed by DEER

The DEER time trace of a 1:1 mixture of HSA and 16-DSA is displayed in Fig. A.4. The exponential decay of the signal originates from homogeneously distributed spins in the sample. Note the absence of a cosine modulation, which precludes a dipolar coupling of two electron spins with a defined distance relationship.

Two conclusions can be drawn from this observation. First, no dimer formation of two HSA molecules is observed. HSA dimers may arise due to the formation of an intermolecular disulfide bond involving the single free cystein residue in the protein [3, 4]. A dimer with a fixed structural relationship between two protein

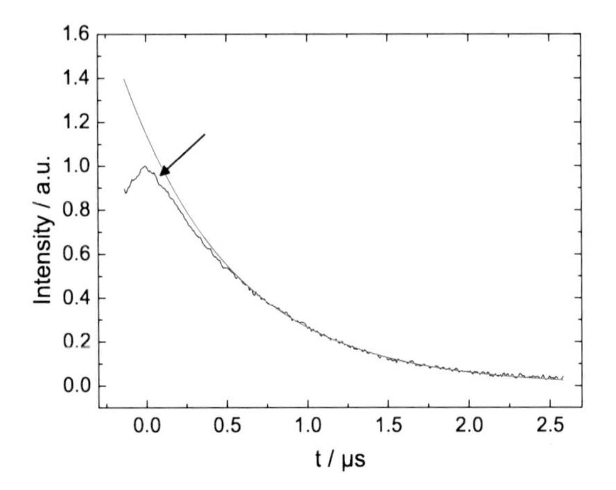

Fig. A.4 Raw DEER data of a HSA–16-DSA mixture with a ratio of 1:1. A fit of the time-domain data with an exponential function $I = A\exp(-bt)$, accounting for homogeneously distributed electron spins in three dimensions, is displayed in *red*. The arrow marks the deviation from the experimental time trace at small times t due to a depopulation of short distances. Reprinted with permission from [12, 13]. Copyright 2010 WILEY-VCH Verlag GmbH & Co. KGaA, Weinheim

molecules would lead to distinct intermolecular distances of the incorporated DSA molecules, which would manifest themselves in dipolar modulations. This conclusion is based on the realistic assumption that at least one intermolecular pair of binding sites should be within a distance range of 6 nm, which can be detected by a dipolar evolution time of 2.7 µs. Even if dimer formation occurred, it would not lead to artifacts in the DEER distance distribution. Second, the fatty acids are distributed equally among all HSA molecules, i.e. one fatty acid molecule is incorporated per protein with a very high selectivity. This indicates a binding site with substantially higher binding affinity for DSA than all other sites. For stearic acid, $K_1 = 9.1 \times 10^8$ and $K_2 = 3.6 \times 10^8$, analyzed by a stepwise equilibrium model [5], are reported which are in agreement with this observation. Note that DSA could exhibit a slightly altered binding affinity due to slight structural deviations.

The deviation from an exponential decay at small dipolar evolution times t originates from a deviation of an ideal homogeneous three-dimensional distribution of all electron spins. Specifically, it is caused by a depopulation of small distances due to an excluded volume effect (proteins cannot overlap with other proteins). This effect is quantified in the next paragraph.

A.1.5 Background Correction

In general, the exponential decay of the DEER signal due to homogeneously distributed electron spins can be described by the function $I = A\exp\left(-bt^{d/3}\right)$, where b is a concentration and relaxation time dependent constant and d is the dimensionality of the sample [6]. For proteins with no preferences for specific

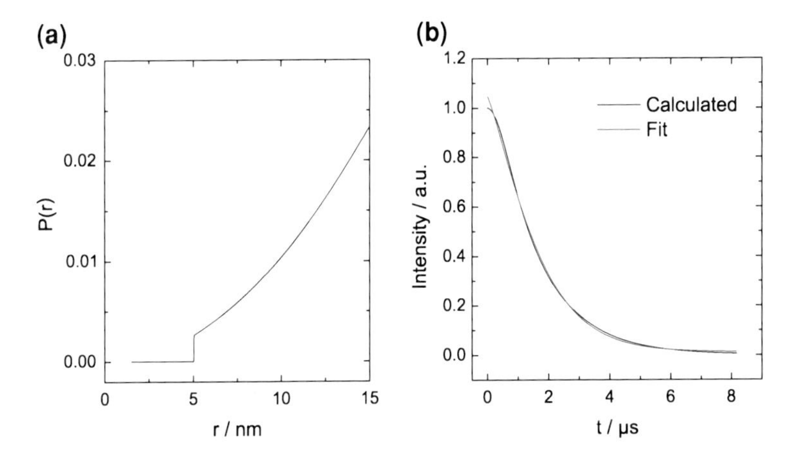

Fig. A.5 **a** Simulated homogeneous distances distribution of electron spins in three dimensions with a lower limit value of 5 nm. **b** Calculated normalized DEER time-domain data (*black*) and fit with an exponentially decaying function (*red*). Reprinted with permission from [12, 13]. Copyright 2010 WILEY-VCH Verlag GmbH & Co. KGaA, Weinheim

surfaces or interfaces (e.g. HSA), $d = 3$ can be assumed. However, the depopulation of short distances due to the excluded volume effect causes a deviation from $d = 3$ towards an apparently higher dimensionality.

For HSA, an approximate diameter of 5 nm was determined from the crystal structure. Hence, 5 nm was taken as lower distance limit for homogeneously distributed electron spins. In this approximation, all electron spins are assumed in the center of a spherical protein molecule. Though not strictly valid, this assumption delivers realistic results regarding the quantification of the expected excluded volume effect. Above the limit of 5 nm, the probability to find other electron spins scales with $4\pi r^2$ for a three dimensional distribution (Fig. A.5a). By this relationship, relative probabilities $P(r)$ of spins within 40 nm (which contribute to the background signal) [7] were determined and a theoretical DEER signal was calculated (Fig. A.5b). By fitting of the signal with an exponentially decaying function, an apparent dimensionality of $d = 3.74$ was obtained.

This calculated value is in excellent agreement with the optimized dimensionality value of the measured DEER data determined by DeerAnalysis2008 program. It ranges from $d = 3.7$ to 4.0 for HSA–DSA complexes of all measured ratios. A value of $d = 3.8$ was consistently used to obtain background corrected data.

A.1.6 Distances Retrieved from Crystal Structure 1e7i

Distance distributions were generated from the crystallographic data to allow for a direct comparison to the fatty acid distance distributions in HSA in solution, as observed by DEER. The distances r_i between the C-16 (or C-5) atoms of the

Table A.1 Distances [nm] between C-16 atoms of stearic acid molecules in HSA obtained from crystal structure 1e7i

	1	2	3	4a	5	6
2	2.23					
3	2.54	4.64				
4a	2.64	4.26	2.13			
5	4.41	6.53	3.17	2.59		
6	3.54	3.83	3.17	3.70	5.90	
7	2.06	3.38	1.76	2.44	4.52	1.32

Table A.2 Distances [nm] between C-5 atoms of stearic acid molecules in HSA obtained from crystal structure 1e7i

	1	2	3	4a	5	6
2	2.48					
3	3.42	3.63				
4a	3.10	3.71	0.89			
5	3.25	5.00	2.68	2.14		
6	4.18	3.29	1.82	2.18	4.31	
7	2.92	1.35	2.59	2.92	4.54	2.18

fatty acids in all sites (1–7) are displayed in Tables A.1 and A.2. Note that the position of the respective carbon atom is an approximation of the unpaired electron position. The displacement from the carbon atom is 0.2 nm, assuming that the electron spin density resides in the middle of N–O bond.

Site 4: Bhattacharya et al. pointed out that the methylene tail of stearic acid traverses the entire width of the hydrophobic tunnel of site 4 formed by the protein. They suggested an alternative binding mode of stearic acids to site 4 in an inverted configuration (4b) [8]. Site 4a is common for all fatty acids regardless of their length and offers additional electrostatic interactions between the carboxylic acid group and Arg410, Tyr411, and Ser489 residues of the protein [8]. Thus, site 4a was chosen for the calculation of all distance distributions. These distance distributions also offer a better agreement with the DEER data.

Sites 1 and 7: In the crystal structure, the stearic acid in binding site 1 is only resolved to its C-13 atom, the fatty acid in site 7 to C-11 [8]. Thus, the distances to the C-16 positions of these sites were extrapolated, assuming that the fatty acids linearly extend along the direction of the five last atoms of their methylene chain. The extrapolated distance values are colored (Table A.1).

A.1.7 Distance Distribution by Consecutive Occupation of Binding Sites

One could assume that the DEER distance distribution upon addition of two FA equivalents to HSA predominantly reflects the distances between the two most tightly bound fatty acids in the high affinity binding sites. In this framework, the addition of more fatty acid equivalents would lead to a broadening of this basic distribution as multiple additional distances are added upon occupation of further sites. Hence, one could try to locate the position of those binding sites with the highest affinity for fatty acids.

The dominant peak in the 16-DSA distribution is centered at 3.6 nm (Fig. 3.3). As can be retrieved by Table A.1, corresponding crystal structure distances are 3.54 nm (distance between sites 1–6), 3.83 nm (sites 2–6), and 3.70 nm (sites 4–6). Note that all potential distances include site 6, which was previously assigned as a low affinity binding site [9]. The corresponding distances of the C-5 positions are 4.18 nm (sites 1–6), 3.29 nm (sites 2–6), and 2.18 nm (sites 4–6). While the latter two distances fit well into the observed distribution, the distance of 4.18 nm between sites 1 and 6 is too high. Hence, this analysis would assign either sites 2 and 6 or sites 4 and 6 as binding sites with the highest affinity for fatty acids.

The analysis can be extended to three bound fatty acids by simulation of the addition of a third equivalent of fatty acid. Let us assume that these fatty acids were predominantly located at binding sites 2, 4, and 6. As demonstrated in the last paragraph, all distances between sites 2 and 6, and 4 and 6 fit in the observed distance distributions. However, the C-16 positions of sites 2 and 4 are at a distance of 4.26 nm, which does not fit to the observed distributions. The same is true for all other possible combinations of three binding sites.

Even at this early stage, the crystallographic distances do not fit to the EPR data. Thus, no meaningful results can be retrieved by this analysis. A conformational change of (parts of) the protein has to be assumed, which renders the initial assumption of this section questionable.

A.1.8 MD Simulations

MD simulations were performed to investigate if the protein may assume a structure that resembles the obtained DEER data better after relaxing in a hypothetical aqueous environment. 1,200 snapshots of the protein structure were

Table A.3 Distances [nm] between C-16 atoms of stearic acid molecules in HSA obtained by MD simulation

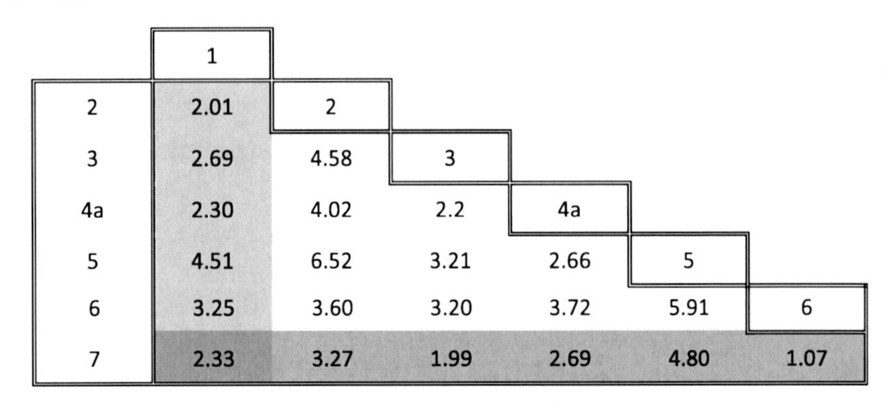

	1	2	3	4a	5	6
2	2.01					
3	2.69	4.58				
4a	2.30	4.02	2.2			
5	4.51	6.52	3.21	2.66		
6	3.25	3.60	3.20	3.72	5.91	
7	2.33	3.27	1.99	2.69	4.80	1.07

Table A.4 Distances [nm] between C-5 atoms of stearic acid molecules in HSA obtained by MD simulation

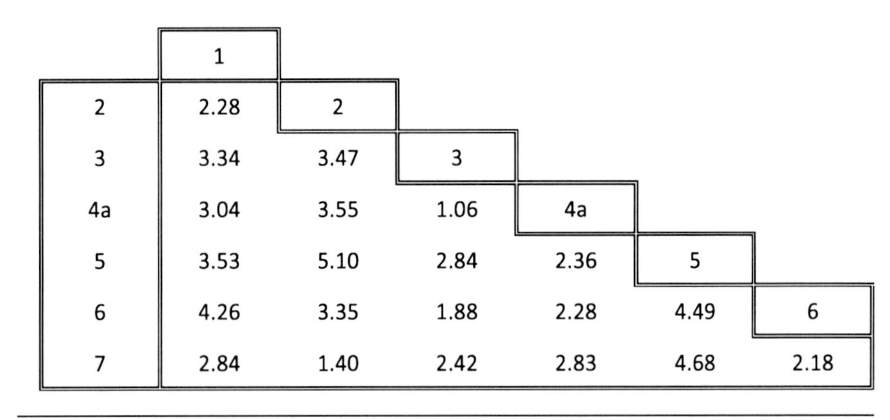

	1	2	3	4a	5	6
2	2.28					
3	3.34	3.47				
4a	3.04	3.55	1.06			
5	3.53	5.10	2.84	2.36		
6	4.26	3.35	1.88	2.28	4.49	
7	2.84	1.40	2.42	2.83	4.68	2.18

averaged and analyzed in terms of distances between complexed fatty acids. The retrieved distances are summarized in Tables A.3 and A.4. Extrapolated distance values to the C-16 positions in sites 1 and 7 are colored (cf. Appendix A.1.6).

The distance distributions obtained by MD simulation exhibit slight but no substantial deviations from those of the original crystal structure (Fig. A.6). For the C-16 position of the fatty acids, the MD structure is in a slightly better agreement with the experimental data than the crystal structure. Especially when contributions from the sparsely resolved fatty acids in the low affinity binding sites 1 and 7 are neglected, a distance distribution is obtained that is similar to the experimental data. Note that distances >6 nm are not accessible by DEER with a dipolar evolution time of 2.5 μs.

For the C-5 position, crystal and MD structure exhibit even minor deviations. Both distributions are in a better agreement with the experimental data in

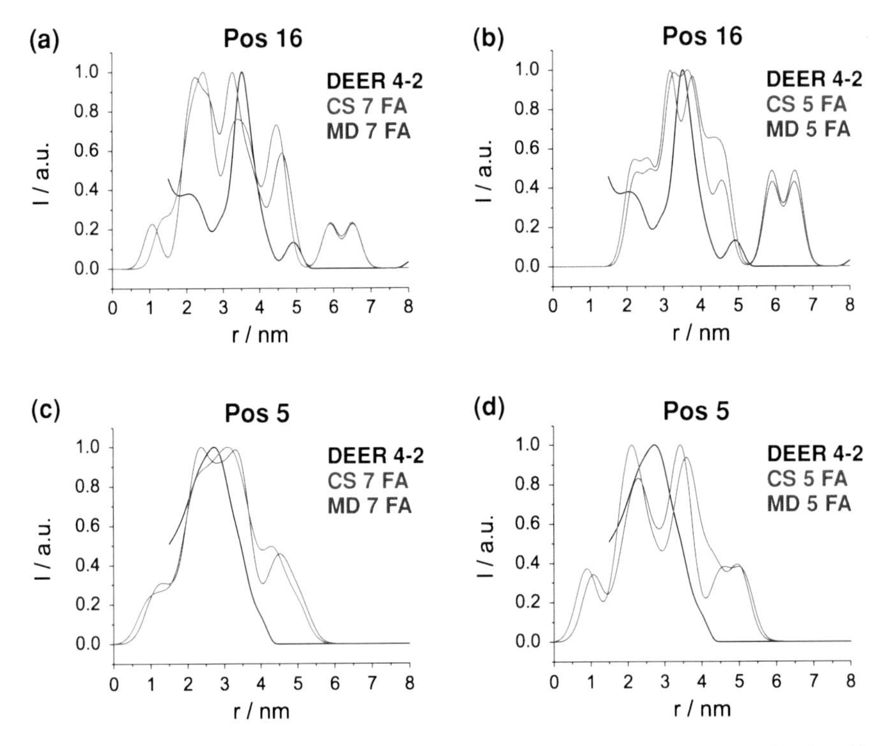

Fig. A.6 Comparison of experimental distance distributions obtained by DEER (*black*) with calculated distributions from the crystal structure (*red*) and MD simulation (*blue*). *Top*: Experimental 16-DSA data and calculated data (position C-16) assuming that **a** all 7 binding sites are occupied by fatty acids and **b** without FA 1 and FA 7. *Bottom*: Experimental calculated data regarding position C-5 of the fatty acid. **c** Occupation of all binding sites, **d** occupation of sites 2–6. Reprinted with permission from [12, 13]. Copyright 2010 WILEY-VCH Verlag GmbH & Co. KGaA, Weinheim

comparison to the C-16 position. Note that the C-5 position is close to the charged anchoring groups of the binding site while the C-16 position is proximate to the entry of the fatty acid channel. Hence, these observations—minor deviations between crystal and MD structure and a better agreement with experimental data in solution—support the suggestion that anchoring region of the protein is rigid while the entry points have a substantial higher degree of freedom and flexibility.

In summary, the MD structure correlates better with the measured DEER data, however changes towards the crystal structure are too small to allow for a thorough conclusion.

A.1.9 ESE Detected Spectra of HSA–16-DSA Mixtures

ESE detected spectra of HSA with varying amounts of 16-DSA are displayed in Fig. A.7. An increasing number of spin-labeled fatty acids per protein molecule

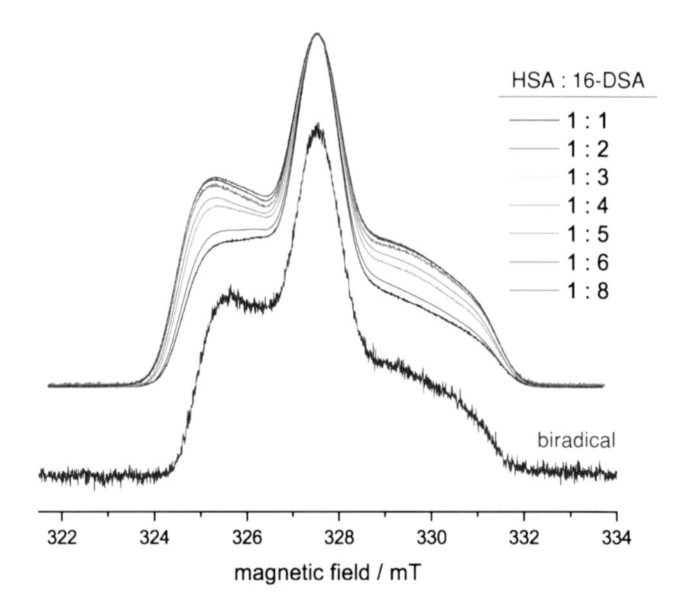

Fig. A.7 ESE detected nitroxide absorption spectra of different HSA–16-DSA mixtures and of a rigid phenylene-ethynylene based biradical with a spin–spin distance of 2.8 nm. All spectra are normalized to their maximum value. Reprinted with permission from [14]. Copyright 2011 Elsevier Inc

leads to an increasing relative height of the spectral flanks, as the transverse relaxation time T_2 of the spin clusters decreases.

Since the spectra are detected by a primary echo sequence, both the echo amplitude and the spectral height depend on the ratio of the interpulse delay τ and the transverse relaxation time T_2. The T_2 time is not homogeneous for the whole spectrum but is slightly higher for regions with strong contributions from g_{zz}. Hence, the central peak is more affected by a reduction of the T_2 time than the spectral flanks.

A.1.10 No Orientation Selection Observed in HSA–DSA Conjugates

The $\sin\theta$ weighting of orientations of the spin–spin vector with respect to the magnetic field (Eq. 2.100) may be skewed, if the orientation of the connecting vector is correlated to the orientation of the molecular frame of at least one of the paramagnetic centers. Such a correlation is not expected for self-assembled DSA molecules in HSA, since the spin-bearing hydrocarbon ring should be able to rotate rather freely around the methylene chain of the fatty acid, even if the fatty acids itself is motionally restricted due to narrow protein channels.

The influence of a possible orientation selection on the DEER signal can be estimated by variation of the observer and pump frequencies within the nitroxide

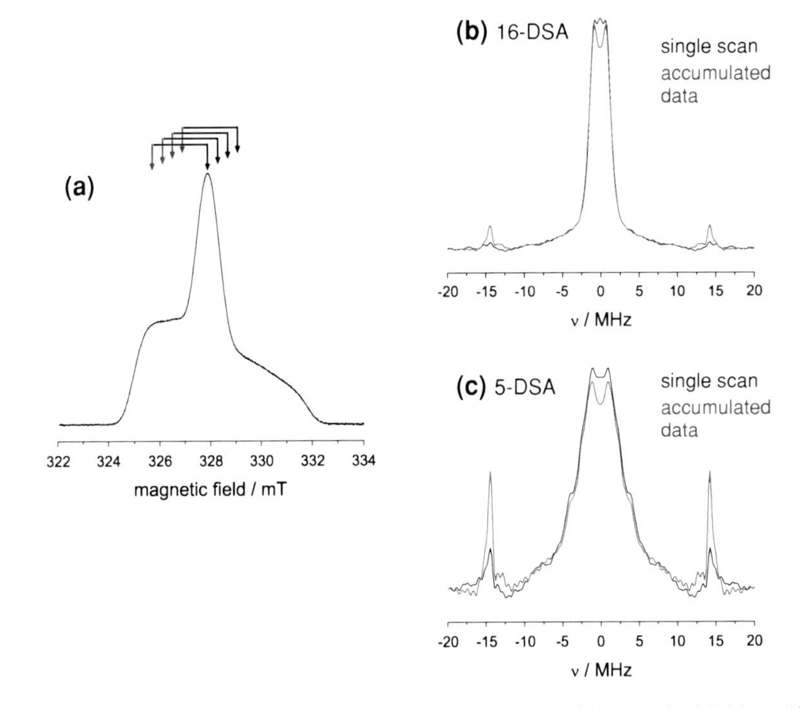

Fig. A.8 **a** ESE detected absorption spectrum of a 1:2 mixture of HSA and 16-DSA at 50 K. Arrows mark magnetic field positions of the observer (*red*) and the pump (*black*) pulse. The positions were varied by a field sweep while the frequency offset was kept constant. 12 equidistant DEER measurements were recorded in a sweep width of 3.6 mT. The dipolar evolution time τ_2 was set to 1,500 ns. **b**, **c** Dipolar spectra of 1:2 complexes of HSA and 16-DSA/5-DSA. Black spectra represent single DEER scans with observer frequencies at the spectral maximum. Spectra in red were obtained by Fourier transformation of accumulated time-domain contributions from all scans. The resonances at ~15 MHz originate from residual ^{1}H ESEEM modulations. Reprinted with permission from [14]. Copyright 2011 Elsevier Inc

spectrum [7]. In a field-swept DEER experiment, the frequency offset between the observer and pump pulses is kept constant, while both frequencies are varied by a sweep of the magnetic field (Fig. A.8a). Thus, potential effects due to orientation selection are averaged and an undistorted Pake doublet is obtained.

The dipolar spectra of HSA admixed with two equivalents of 5-DSA and 16-DSA are shown in Fig. A.8 b, c. The spectrum in black originates from a single DEER scan with the pump pulse at the maximum spectral position while accumulated data from the whole range of the field sweep is displayed in red. The spectra exhibit no significant deviations. Hence, no angular correlations are observed for both 5-DSA and 16-DSA.

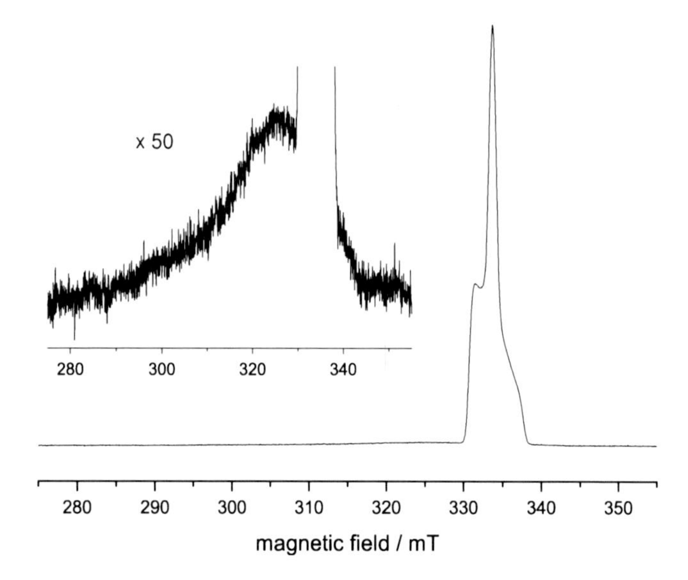

Fig. A.9 ESE detected spectrum of HSA admixed with 1 eq. hemin and 1 eq. 16-DSA. In the inset, the spectrum is magnified by a factor of 50 to visualize the spectral contribution from Fe^{3+}. Reprinted with permission from [15]. Copyright 2011 Biophysical Society

A.1.11 ESE Spectrum of HSA Complexed With Hemin and 16-DSA

Hemin contains a paramagnetic center that gives rise to an EPR signal. However, Fe^{3+} possesses a quadrupolar electron spin with $S = 5/2$, which leads to a substantial broadening of the EPR spectrum (Fig. A.9). Due to the low spectral density, DEER spectroscopy between the iron center and the nitroxide suffers from a low modulation depth and from a bad SNR independent of the spectral position of the pump pulse. Hence, hemin was replaced by the Cu(II) analogue with an electron spin of $S = 1/2$.

A.1.12 Copper–Nitroxide DEER Without Spin Dilution

In Fig. A.10, orientation-selective dipolar data between Cu(II) protoporphyrin IX and a varying number of paramagnetic 16-DSA molecules are displayed. Higher fractions of spin-labeled fatty acids lead to increased modulation depths in analogy to the nitroxide–nitroxide DEER measurements in Sect. 3.3. The dipolar spectra exhibit a decrease of the dominant frequency at ~ 1 MHz, while strong dipolar contributions become more prominent at the same time. This trend is in full analogy to the results in Sect. 3.3. Large dipolar couplings are overestimated due to multispin interactions while small dipolar couplings are suppressed.

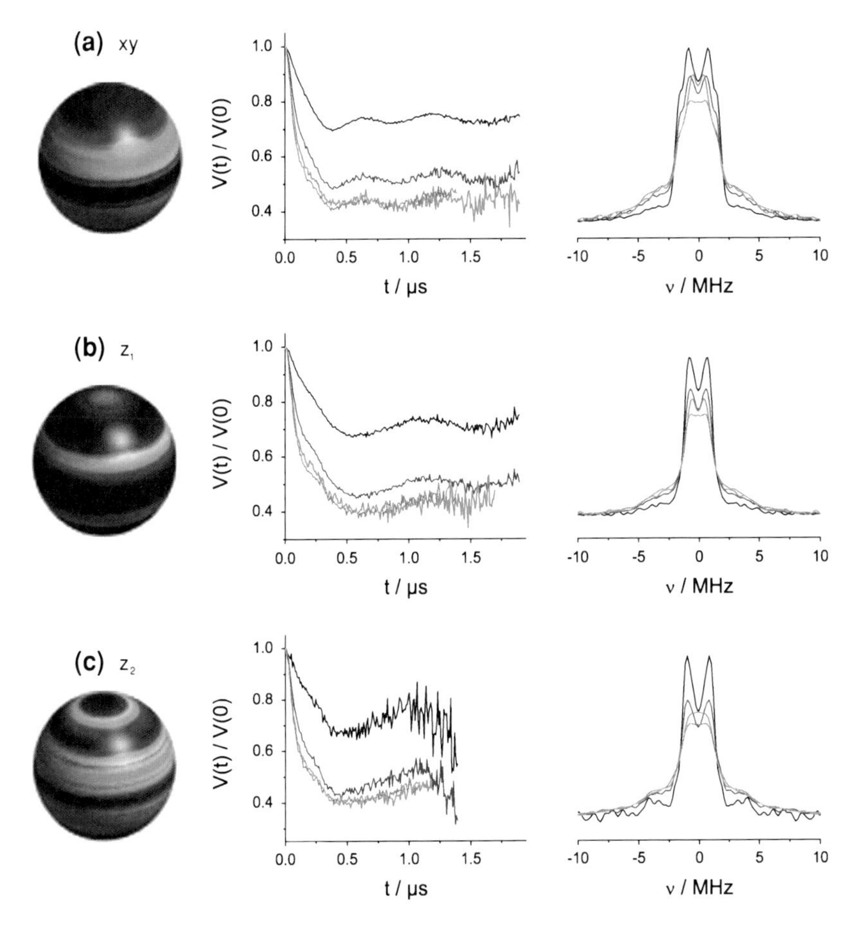

Fig. A.10 Background-corrected DEER time-domain data and dipolar spectra of HSA complexed with 1 eq. of Cu(II) protoporphyrin IX and varying equivalents of 16-DSA. The observer pulse was applied at field positions exciting Cu(II) orientations predominantly in the *xy*-plane of the porphyrin ring (**a**, position xy) and outerdiagonal elements towards the molecular *z*-axis (**b**, position z_1; and **c**, position z_2). Reprinted with permission from [15]. Copyright 2011 Biophysical Society

A.1.13 Simulation Program for Orientation Dependent DEER Spectra

In the following, the Matlab source code of the DEER simulation program is presented. Explanations are displayed in green.

```
%% orientation selection via orisel
g_Cu=[2.0530 2.194];
Exp=struct('Range',[250 350],'nPoints',2048,'mwFreq',9.280497,
'Harmonic',1);
opt=struct('Verbosity',1,'nKnots', 100);
Sys1=struct('S',1/2,'g',g_Cu,'HStrain',[20 20 45],'lw',0.5);
Sys2=struct('S',1/2,'g',g_Cu,'HStrain',[20 20 45],'lw',0.5);
Sys1.Nucs='63Cu,14N,14N,14N,14N';Sys1.A=mt2mhz([2.10 22.0; 1.8 1.35;1.8
1.35;1.8 1.35;1.8 1.35]);
Sys2.Nucs='65Cu,14N,14N,14N,14N';Sys2.A=mt2mhz([2.10*1.588/1.484
22.0*1.588/1.484; 1.8 1.35;1.8 1.35;1.8 1.35;1.8 1.35]);

Exp.Field=335.5; %field position of observer pulse
Exp.ExciteWidth=31.25; % excitation bandwidth of observer pulse = 1/tp

TotalWeights1 = orisel(Sys1,Exp,opt);
TotalWeights2 = orisel(Sys2,Exp,opt);
TotalWeights = 0.6917*TotalWeights1+0.3083*TotalWeights2;
B = TotalWeights/max(TotalWeights);

%% orientation-dependent DEER calculations
%define phi and theta axes
phi=linspace(0,2*pi-pi/opt.nKnots,2*opt.nKnots);
theta=linspace(0,pi,2*opt.nKnots-1);

for y=2:length(B),
 Bneg(y-1)=B(length(B)+1-y);
end;
Ba=[B;Bneg'];

% distances/angles from crystal structure
r=[2.36 2.98 2.38 4.25 3.82 2.67]; % Cu to C-16 distances
thetad=[64.7 56.2 85.7 73.2 88.5 75.8]*pi/180; %angles of connection
vectors with respect to z-axis of Cu

%definition of other parameters for calculation
gN=2.00581; %g-value for nitrogen assumed isotropic
ge=2.0023; %g-value of the free electron
np2=401; % number of points in dipolar and time dimension

swdip=24; % width of dipolar spectrum

%start parameters
dip0=zeros(1,np2);
dip1=dip0;
deer_frq=dip0;
deer_frq2=dip0;
deer_time=dip0;
deer_time3=dip0;
dipax=linspace(-swdip/2,swdip/2,np2);
time=linspace(0,8,np2);
sr1=0.2;
%% Gaussian distance peaks
for y=1:length(r),
  hom=0.1;
  nax=40; % numbers of points of r-axis, nax=40 => dr=0.05
```

```
distr=zeros(1,nax);
rax=linspace(r(y)-1,r(y)+1,nax); % definition of r-axis +/-1 nm around
peak maximum
dist1=(rax-r(y))/sr1;
distr=distr+hom*exp(-dist1.*dist1); % Gaussian distribution
distr=distr/max(distr); %normalization to peak maximum

for z=1:length(rax),
 nydd=52.04/rax(z)^3; % dipolar frequency [MHz/nm^3]
 ddvec=[sin(thetad(y)) 0 cos(thetad(y))]; % dipole-dipole vector in the
 reference frame of spin 1

 for k=1:length(phi), % integration over all angles phi
  for n=1:length(theta), %integration over all angles theta
   st=sin(theta(n));
   ct=cos(theta(n));
   cfi=cos(phi(k));
   sfi=sin(phi(k));
   weight=Ba(n); %weighting of orientations according to orisel
   if weight>1e-2, % lower limit for orientational contributions
    if distr(z)>5e-2, %lower limit for contributions from r-values (5% of
    peak maximum)
     B0vec=[st*cfi,st*sfi,ct]; % magnetic field unit vector in frame of
     Cu(II) porphyrin
     cthdd=sum(B0vec.*ddvec); % cosine of angle between dipole-dipole
     vector and magnetic field
     gCu=sqrt((g_Cu(1))^2*st^2+(g_Cu(2))^2*ct^2); % g-value of Cu at the
     respective orientation
     dipfrq=gCu*gN/ge^2*nydd*(3*cthdd^2-1); % dipolar frequency at this
     orientation
     dip1=put(dip0,swdip,dipfrq,1); % create positive part of dipolar
     doublet
     dip1=put(dip1,swdip,-dipfrq,1); % create negative part of dipolar
     doublet
     deer_frq=deer_frq+st*weight*distr(z)*dip1; % weighting of frequency
     spectrum with sin(theta)
     deer_time=deer_time+st*weight*distr(z)*(1-cos(2*pi*dipfrq*time)); %
     time-domain data
          else
          end;
         else
         end;
        end;
       end;
      end;
     end;
```

The dipolar spectrum is constructed by the function 'put', which was kindly provided by G. Jeschke:

```
function spc=put(spc,sw,frq,weight)

np=length(spc); % number of data points
step=sw/(np-1); % frequency interval
nya=-sw/2; % starting frequency (relative to center)
poi=1+round((np-1)*(frq-nya)/sw); % data point in spectrum
dny=frq-nya-(poi-1)*step; % frequency difference from center point
delny=linspace(-3*step-dny,3*step-dny,7)/5;
eline=exp(-delny.^2);
if poi>3 && poi<=np-3,
    spc(poi-3:poi+3)=spc(poi-3:poi+3)+weight*eline;
end;
```

A.2 Copper Complexes of Star-Shaped Cholic Acid Oligomers with 1,2,3-Triazole Moieties

A.2.1 *Spectral Isolation of 'Bound' and 'Intermediate' Species*

Different ratios of 'bound' and 'intermediate' copper species are present in the EPR spectra dependent on the relative amount of T3t that is admixed to $CuCl_2$. Thus, a spectral separation of both species is achieved by linear combinations of the recorded spectra (Fig. A.11, black curves). These isolated spectral components allowed for unpretentious simulations (red curves), which were then combined to match the original multi-component spectra (Fig. 4.3).

A.2.2 *Comparison of T3t and T3b as Pyrene Hosts*

The intensity ratio of the first and third peak in its fluorescence spectrum, I_3/I_1, serves as a measure for the polarity of the immediate environment around a pyrene molecule, since it is proportional to the hydrophobicity of the solvent. In the presence of T3t and T3b, the ratio I_3/I_1 increases with an increasing concentration of the host (Fig. A.12). This indicates that pyrene enters a less polar environment due to its inclusion in the hydrophobic cavities.

At the same concentration, the ratio I_3/I_1 is slightly decreased, when T3b is added as a host instead of T3t. A ratio is further decreased when 1 eq. Cu^{2+} was added to T3b prior to pyrene. This indicates that pyrene is not as readily accommodated in the hydrophobic cavity of T3b as in the pocket formed by T3t, especially after the addition of a metal ion. However, the observed effects are small and cannot account for the significant difference in the fluorescence quenching efficiencies of T3t and T3b.

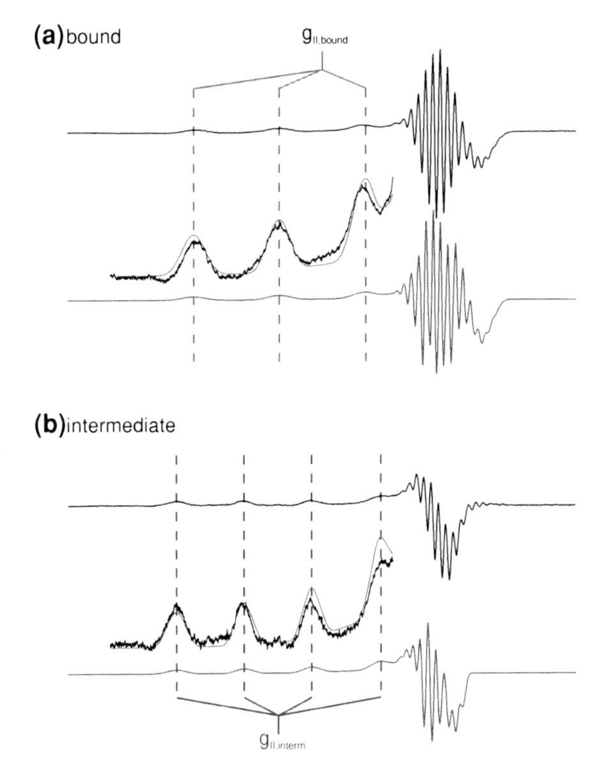

Fig. A.11 Isolated spectral contributions of 'bound' and 'intermediate' copper species in $CuCl_2$–T3t spectra recorded at 78 K. Spectra are shown in *black*, the corresponding simulations in *red*. The low field regions are magnified by a factor of ~ 15. The spectra and simulations were normalized with respect to their double integral. The peak positions of the spectral components parallel to the unique axis of the distorted octahedral Cu(II) frame are indicated by *orange* (bound) and *blue* (intermediate) bars. The isolated spectral contributions R were obtained by linear combinations of the recorded spectra of Fig. 4.3: $R(\text{bound}) = \text{T3t}(1:2) - 0.56\ \text{T3t}(1:1) + 0.056\ \text{T3t}(1:0)$, and $R(\text{interm.}) = \text{T3t}(1:1) - 0.68\ \text{T3t}(1:2) - 0.1\ \text{T3t}(1:0)$

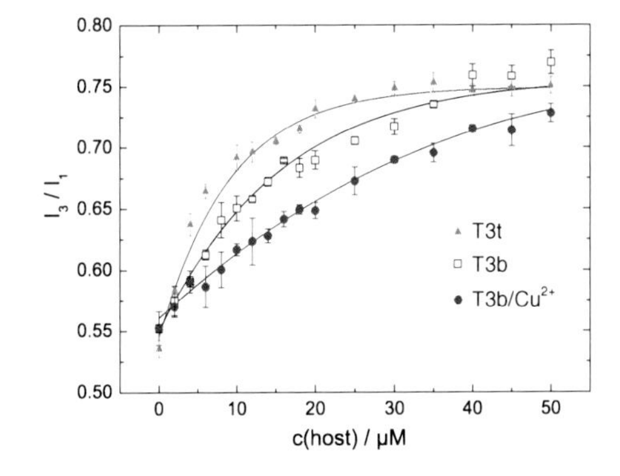

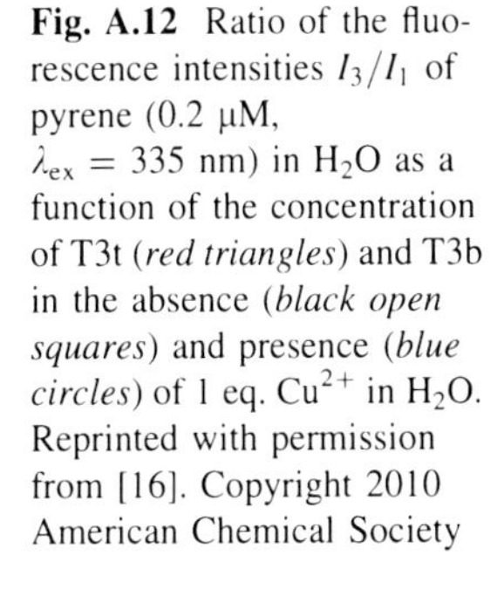

Fig. A.12 Ratio of the fluorescence intensities I_3/I_1 of pyrene (0.2 μM, $\lambda_{ex} = 335$ nm) in H_2O as a function of the concentration of T3t (*red triangles*) and T3b in the absence (*black open squares*) and presence (*blue circles*) of 1 eq. Cu^{2+} in H_2O. Reprinted with permission from [16]. Copyright 2010 American Chemical Society

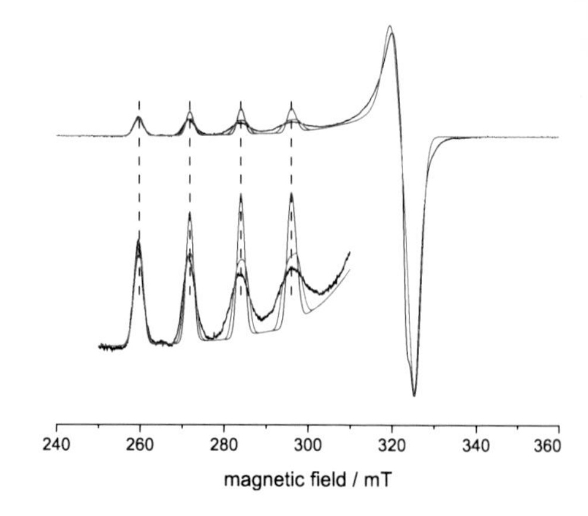

Fig. A.13 Spectrum of $CuCl_2$ in a ternary solvent mixture of ethanol, water, and glycerol (9:1:2 v/v) at 78 K (*black*) and corresponding simulations (*blue*, *red*), as specified in the main text. The low field regions are magnified by a factor of ~15. Both spectrum and simulations are normalized to their double integral

A.2.3 *Origin of the Inhomogeneous Broadening of Non-Coordinated $CuCl_2$*

For the 'free' Cu^{2+} species, and to a less extent for the both 'loosely bound' species, the hyperfine lines in the $g_{\|}$ region are broadened towards a higher field (Fig. A.13). This broadening cannot be accounted for by unresolved hyperfine couplings or by a strains of the **g** or **A** tensors, as these broadenings would equally affect all hyperfine transitions.

It was found that, in a first approximation, the deviation of $g_{\|}$ from g_e due to spin–orbit coupling is antiproportional to $A_{\|}$, i.e.

$$\left(g_{\|} - g_e\right)A_{\|} \cong \text{const.} \tag{A.1}$$

In fact, values between 163.3 and 168.0 MHz were obtained for all species in this study, with the exception of the 'bound' species, which exhibits a slightly lower value of 148.7 MHz. Although the increase of $A_{\|}$ does not fully account for the decrease of $\Delta g_{\|}$, both values are strongly coupled. Hence, a strain of the **g** tensor induces a coupled strain of the **A** tensor.

In Fig. A.13, two spectral simulations of the free Cu^{2+} species are shown. The best fit of a simulation with a single Cu^{2+} species with $g_{\|} = 2.419$ and $A_{\|} =$ 393.7 MHz is depicted in blue. In the second simulation, displayed in red, three different species equally contribute to the overall spectrum. These species exhibit slightly different $g_{\|}$ and $A_{\|}$ values, which were coupled according to Eq. A.1. Specifically, data pairs $(g_{\|}, A_{\|}) = (2.429, 393.7\ \text{MHz})$, (2.419, 403.2 MHz), and (2.409, 413.7 MHz) were chosen, so that $\left(g_{\|} - g_e\right)A_{\|} = 168$ MHz was fulfilled for all fractions. Since the main focus was on the low field region, the EPR parameters perpendicular to the unique octahedral axis were kept constant for both simulations at $g_{\perp} = 2.805$ and $A_{\perp} = 28$ MHz.

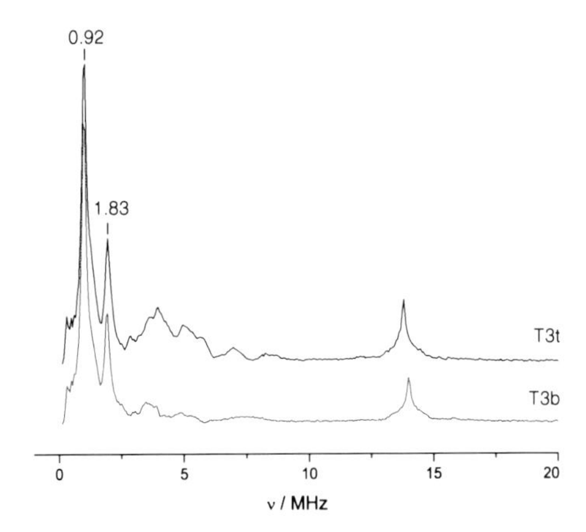

Fig. A.14 3-Pulse ESEEM spectra of 1:2 mixtures of $CuCl_2$–T3t and $CuCl_2$–T3b at the EPR spectral maximum (corresponding to $g_\perp$). The displayed spectra are the average of three scans with τ values of 132, 164, and 200 ns. All spectra were recorded at 10 K

As can be clearly seen in the magnification of the spectrum, the latter approach mimics the asymmetric broadening of the hyperfine lines. The inherently coupled g and A distributions induce a significant broadening of the fourth low field transition, while both effects cancel out each other for the first hyperfine line.

In a more sophisticated approach, a Gaussian distribution of $g_\parallel$ and $A_\parallel$ values could be assumed. The width of this distribution would then provide quantitative information about ligand field inhomogeneities of the copper complexes. Even though this approach was not chosen, qualitative conclusions can be deduced from the spectral data. The largest relative broadening is observed for free $CuCl_2$ in the ternary solvent mixture, since it forms various complexes with a large variety of ligands (water, ethanol, glycerol, and chloride). In contrast, no comparable broadening is observed for the 'bound' and 'intermediate' species. Hence, the copper ions are located in a well-defined ligand field due to the presence of strong, chelating ligands. An intermediate broadening is observed for both 'loosely bound species'. Here, chelating groups are absent, which provide for a defined coordination sphere. As the terminology suggests, the Cu^{2+} ions are rather loosely bound to the triazole groups and can form a variety of complexes with different ligand fields.

A.2.4 3-Pulse ESEEM and Supplementary HYSCORE Data of T3b and T3t

3-Pulse ESEEM spectra of $CuCl_2$ admixed with 2 eq. T3t and T3b are displayed in Fig. A.14. Both spectra exhibit sharp dominant features at 0.92 and 1.83 MHz, which can be assigned to nuclear quadrupole transitions ν_- and ν_+ of remote nitrogen atoms in the limit of exact cancellation [10]. In contrast, major deviations

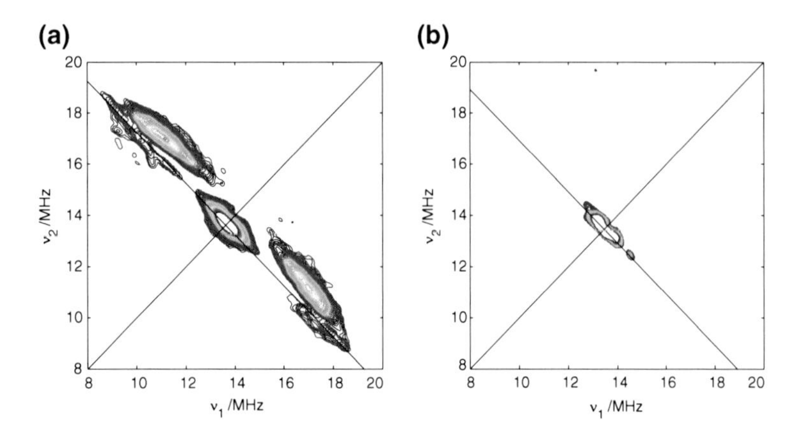

Fig. A.15 ^{1}H region of the HYSCORE spectra of a 1:2 mixture of $CuCl_2$ and T3t in the **a** protonated and **b** deuterated ternary solvent mixture ethanol/water/glycerol (9:1:2 v/v). The pulses were applied at the maximum of the Cu(II) spectrum. The spectra were recorded at 10 K, τ was set to 132 ns. The antidiagonal characterizes the ^{1}H frequency at the applied magnetic field

are observed in the frequency regime of 2.5–10 MHz. $CuCl_2$ in T3t exhibits a multitude of broad features which are absent for $CuCl_2$ in T3b. Spectra similar to T3b were obtained for S2. Regarding the structural differences of T3t and T3b, they are assigned to multiquantum transitions that originate from a multitude of remote ^{14}N atoms in the triazole unit.

In addition to the detailed Cu(II)–triazole binding studies, it was probed if other parts of the cholic acid oligomers act as ligands for the Cu^{2+} ion. In Fig. A.15a, the proton region of the HYSCORE spectrum of a 1:2 mixture of $CuCl_2$ and T3t in a protonated solvent mixture is depicted. No substantial differences were observed for all cholic acid oligo- and monomers in protonated solvents.

Besides the proton matrix peak, which originates from remote hydrogens, two types of strongly coupled protons are observed. Scalar coupled protons exhibit resonances along the antidiagonal, while dipolar coupled protons give rise to the observed ridge. The division of the ridge into two parts is a blindspot artifact. The missing inner part could be retrieved by setting τ to 164 ns.

All resonances from strongly coupled protons are absent when the sample is placed in fully deuterated solvents (Fig. A.15b). Thus, the coupling to protons solely originates from solvent molecules, which bind to the Cu^{2+} ion by their oxygen atoms.

Table A.5 Spectral Contribution the Observed Copper Species for Different Cu–S2 Ratios

Ratio $CuCl_2$: S2	Relative spectral contribution		
	Free	Loosely bound1	Loosely bound2
2:1	0.61 ± 0.03	0.27 ± 0.04	0.12 ± 0.04
1:1	0.21 ± 0.03	0.48 ± 0.08	0.31 ± 0.08
2:3	0.17 ± 0.03	0.44 ± 0.08	0.39 ± 0.08
1:2	0.13 ± 0.03	0.34 ± 0.09	0.53 ± 0.09
1:3	0.07 ± 0.03	0.22 ± 0.09	0.71 ± 0.09
1:4	0.04 ± 0.03	<0.05	>0.92

A.2.5 Relative Spectral Contributions in Mixtures of $CuCl_2$ and S2

See Table A.5.

A.3 Chemical Nitroxide Decay in Thermoresponsive Hydrogels

A.3.1 Spin Probe Decay in Hydrogels P1 and P2

At 5 °C, gel P1, swollen with pure water, and gels P1 and P2, swollen with 0.05 M HCl, yield comparable catalytic activities (Fig. A.16a). In contrast, the catalytic activity of the hydrogel solutions containing HCl is strongly accelerated at 50 °C, while it is considerably decreased in a pure aqueous solution of the collapsed crosslinked polymer P1 (Fig. A.16b).

The anomalous feature of a decreasing catalytic activity with increasing temperature is not observed when P1 is swollen with a buffer solution instead of pure water (Fig. A.16c, d). These buffer solutions exhibit an enhanced catalytic activity at 50 °C since the localized carboxylic acid protons of P1 are mainly transferred to the mobile phosphate buffer and not affected by an altered hydrogen bonding framework.

A.3.2 Reference Data Sets

Neither an aqueous solution of non-crosslinked polymer P1 nor a buffer solution of gel P2 cause a significant enhancement of the nitroxide decay, since they both lack one key factor that triggers the catalytic activity (Fig. A.17). In non-crosslinked P1, a hydrogel network is absent that provides for the spatial confinement and for a large interfacial area between heterogeneities. Gel P2 fulfills this requirement, but it lacks acidic protons, which initiate and catalyze the chemical disportionation.

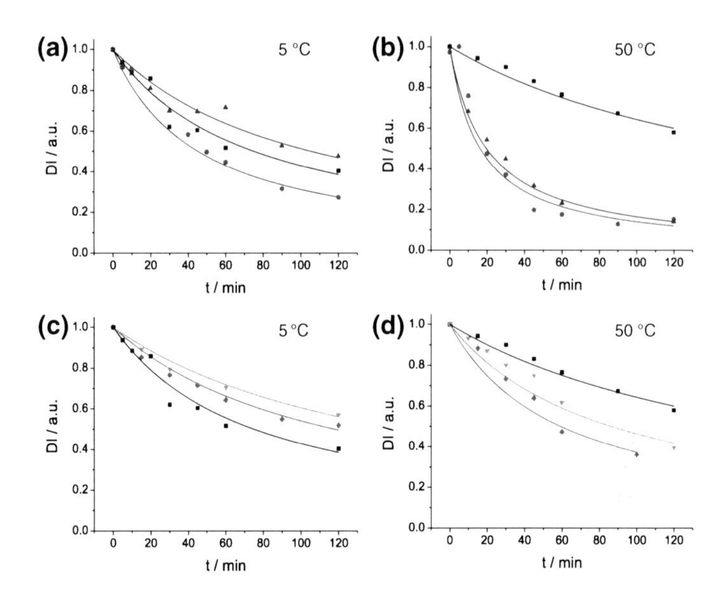

Fig. A.16 Nitroxide decay in hydrogels P1 and P2 at various solvent conditions and temperatures. **a**, **b** Influence of the availability of mobile acidic protons: (*filled square*) P1 gel in pure water, (*filled circle*) P1 gel + 0.05 M HCl, and (*filled triangle*) P2 gel + 0.05 M HCl. **c**, **d** Catalytic activity of P1 in the presence of buffer solutions: (*filled square*) pure MilliQ water, (*filled down pointing triangle*) 25 mM phosphate buffer (pH 7.4), (*filled diamond*) 330 mM phosphate buffer (pH 7.4). The corresponding EPR spectra were recorded at 5 °C (**a**, **c**), and at 50 °C (**b**, **d**). Reprinted with permission from [17]. Copyright 2008 WILEY-VCH Verlag GmbH & Co. KGaA, Weinheim

Despite the fact that a disproportionation was mentioned as only possibility for spin probe degradation, a redox reaction of TEMPO with the hydrogel could also be responsible for the decay. Amide and isopropyl groups in the NiPAAm unit might be potential reaction partners. To shine light onto this question, spectra of TEMPO in gel P3 was recorded since P3 exhibits a collapse behavior comparable to gel P1, but does neither contain NH- nor isopropyl groups. A catalytic enhancement similar to gel P1 was observed (Fig. A.17). Hence, a possible direct participation of amide protons and isopropyl groups in the reaction could be excluded.

A.3.3 Reaction Rate Constants

Reaction rate constants k of the nitroxide decay were obtained by fitting of the measured data sets with the function $\mathrm{DI} = (kt + 1)^{-1}$, assuming a second order reaction decay. The double integrals were normalized to $t = 0$. The extracted reaction rate constants are displayed in Table A.6.

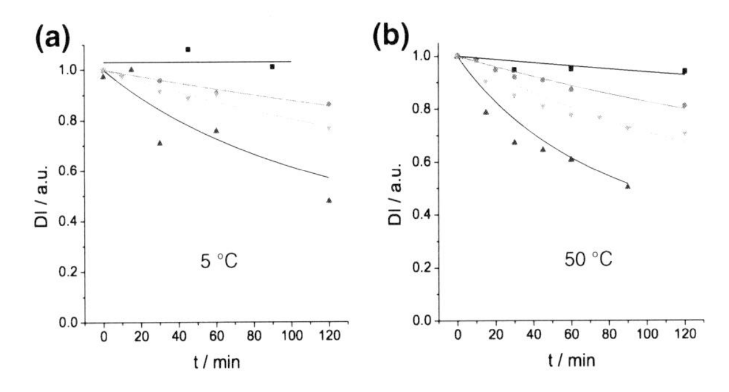

Fig. A.17 Nitroxide decay in different aqueous media at 5 and 50 °C: (*filled square*) pure MilliQ water, (*filled circle*) P2 gel (25 mM buffer), (*filled down pointing triangle*) P1, non-crosslinked (10 wt% in pure water), and (*filled triangle*) P3 gel (25 mM buffer). Reprinted with permission from [17]. Copyright 2008 WILEY-VCH Verlag GmbH & Co. KGaA, Weinheim

Table A.6 Reaction Rate Constants of the Nitroxide Decay under all Studied Conditions

	5 °C	50 °C
	10^3 k/min^{-1}	10^3 k/min^{-1}
25 mM phosphate buffer	0	0.65 ± 0.17
0.05 M HCl	1.55 ± 0.08	4.28 ± 0.47
P1, non-crosslinked	2.45 ± 0.14	4.05 ± 0.24
P1, pure water	13.3 ± 0.8	5.56 ± 0.24
P1, 25 mM buffer	6.53 ± 0.29	11.6 ± 0.8
P1, 330 mM buffer	8.49 ± 0.33	16.9 ± 1.0
P1 + 0.05 M HCl	22.2 ± 0.7	61.7 ± 5.8
P2, 25 mM buffer	1.41 ± 0.08	2.09 ± 0.11
P2 + 0.05 M HCl	9.37 ± 0.42	51.1 ± 1.3
P3, 25 mM buffer	6.32 ± 0.73	10.4 ± 0.5

A.4 Local Nanoscopic Heterogeneities in Thermoresponsive Dendronized Polymers

A.4.1 Detailed Description of the Spectral Analysis

All reported nitroxide parameters in Chap. 7 were obtained by simulations that were fitted to the CW EPR spectra. In Fig. A.18, EPR spectra at various temperatures (10 wt% PG1(ET)) are superimposed by their corresponding simulations. Below T_C, the simulations consist of a single spectral component C with isotropic g- and hyperfine coupling values, g_C and a_C. Above T_C, two spectral components A and B with parameters $g_{A/B}$ and $a_{A/B}$ can be distinguished. In this case, the actual simulation S_{tot} is a linear combination of the simulations for species A (S_A)

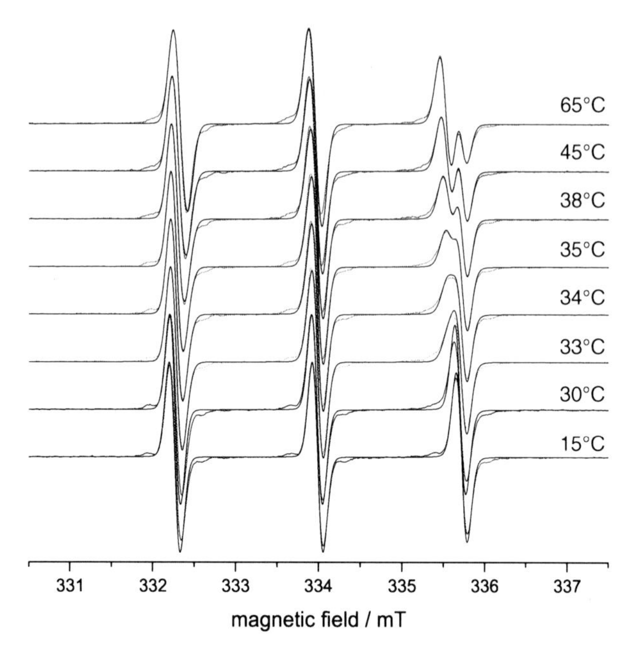

Fig. A.18 CW EPR spectra of 0.2 mM TEMPO in a 10 wt% PG1(ET) aqueous solution at selected temperatures. The corresponding spectral simulations are superimposed in *black*. At 15 and 30 °C, one single spectral component C was assumed. At $T \geq 33$°C, the spectral simulations consist of two distinct components, A and B. Reprinted with permission from [18, 19]. Copyright 2010 WILEY-VCH Verlag GmbH & Co. KGaA, Weinheim

and species B (S_B), as described by Eq. 5.1. In Fig. A.20, the spectral weights x_A are plotted against the temperature.

It is difficult to obtain precise g_{iso} values for spectra, which consist of two overlapping components. However, the correct determination of g_{iso} is crucial since a_{iso} and g_{iso} are coupled and affect each other.[1]

In Fig. A.19b, the relation of a_{iso} and g_{iso} is displayed. This graph was obtained by recording a 0.2 mM TEMPO solution in pure water. The spin probe, located solely in a hydrophilic environment, gave rise to a single spectral species, which could be easily analyzed, i.e. simulated in terms of a_{iso} and g_{iso}. Different data points originate from measurements at various temperatures. Note that water provides a more hydrophobic environment with increasing temperature since its dielectricity constant decreases. This gives rise to a slight decrease of a_{aq}, which is displayed in Fig. A.19a.

Since a_A and a_C are in the range of the measured a_{aq} values (48.2–48.65 MHz), the corresponding g-value could be graphically retrieved from Fig. A.19b. In contrast, the

[1] Other simulation parameters like the linewidth and the rotational correlation time affect the weights x_i of the spectral components. Besides g_{iso} and a_{iso}, all other parameters were thus kept constant for a set of spectra at different temperatures to allow for comparability.

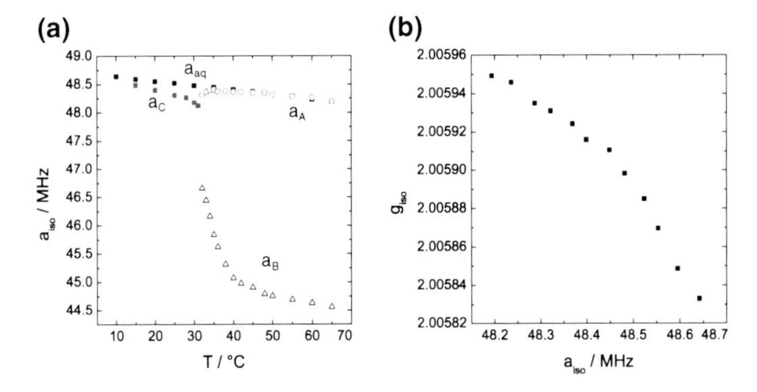

Fig. A.19 **a** Hyperfine coupling values for TEMPO in 10 wt% PG1(ET). Below $T_C = 33$ °C, a single spectral component C with an isotropic hyperfine coupling a_C (*purple*) is observed. Above T_C, two spectral components A and B give rise to hyperfine couplings a_A (*red*) and a_B (*blue*). For comparison, hyperfine couplings of TEMPO in pure water a_{aq} (*black*) are depicted in the same temperature range. **b** Relation of g_{iso} and a_{iso} as determined by analysis of TEMPO spectra in pure water at temperatures from 10 to 50 °C. Reprinted with permission from [18, 19]. Copyright 2010 WILEY-VCH Verlag GmbH & Co. KGaA, Weinheim

coupling values of the hydrophobic spin probe a_B deviate significantly from this region at all temperatures (44.5–47.0 MHz). Thus, the corresponding g_B-values were obtained by linear extrapolation of the data in Fig. A.19b.

The obtained values a_B are *effective* hyperfine coupling values which originate from the exchange of the spin probe B between a hydrophilic and a hydrophobic environment. If one assumes that the static hyperfine coupling constant of spin probes solely located in a hydrophobic region is given by a_B (65 °C), the effective coupling constant a_B is given by

$$a_B(T) = a_B(65\ ^\circ\mathrm{C}) + \chi\big(a_{aq}(T) - a_B(65\ ^\circ\mathrm{C})\big). \tag{A.2}$$

χ quantifies the fraction of hydrophilic regions that contribute to a_B. Likewise, any deviation of a_A from a_{aq} (apparent at temperatures close to T_C) is interpreted in terms of a slight hydrophobic contribution.

Hence, the overall fraction of spin probes in a hydrophilic environment y_A is not represented by the weight x_A of its spectral component (as e.g. in Chap. 5). At $T \geq T_C$ it is calculated by

$$y_A(T) = x_A \frac{a_A(T) - a_B(65\,^\circ\mathrm{C})}{a_{aq}(T) - a_B(65\,^\circ\mathrm{C})} + (1 - x_A)\frac{a_B(T) - a_B(65\,^\circ\mathrm{C})}{a_{aq}(T) - a_B(65\,^\circ\mathrm{C})}. \tag{A.3}$$

At temperatures below T_C, where only one spectral species is observed, y_A amounts to

$$y_A(T) = \frac{a_C(T) - a_B(65\,^\circ\mathrm{C})}{a_{aq}(T) - a_B(65\,^\circ\mathrm{C})}. \tag{A.4}$$

In Fig. A.20, the spectral weights x_A and the total weights y_A of the hydrophilic species are plotted as a function of temperature.

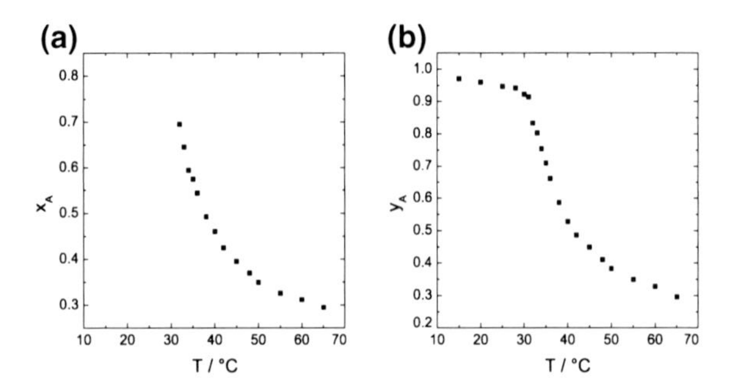

Fig. A.20 Spectral parameters for TEMPO in 10 wt% PG1(ET). **a** Plot of the spectral weight x_A of the hydrophilic species as a function of temperature ($T \geq T_C$). **b** Depiction of the total fraction y_A of TEMPO in a hydrophilic environment as a function of temperature. Reprinted with permission from [18, 19]. Copyright 2010 WILEY-VCH Verlag GmbH & Co. KGaA, Weinheim

A.4.2 EPR and Turbidity Collapse Curves

In Fig. A.21, the collapse curves derived by EPR spectroscopy and turbidity measurements are compared. While turbidity measurements utilize the scattering of light to characterize the aggregation of the polymer, EPR spectroscopy measures the distribution of hydrophilic and hydrophobic regions and thus characterizes the dehydration of polymer segments due to the loss of water.

The first pronounced drop of y_A defines the critical temperature T_C derived from EPR. This is the lowest temperature, at which two separate contributions to the EPR signal can be recognized. Though the obtained EPR curves are very broad and extent to temperatures far beyond T_C, this first drop is well-defined ($\Delta T_C = 1\text{K}$).

In comparison to the critical temperature obtained by turbidity measurements, the slightly lower EPR-derived T_C values originate from higher concentrations of the polymer solutions (10 wt% (EPR) vs 0.25 wt% (turbidity)). Turbidity measurements on comparable dendronized polymer systems showed that the aggregation temperature between highly diluted and concentrated solutions differs by approximately 2 K, which is in good agreement with the observed deviations (Fig. A.21).

A.4.3 Estimation of the Size of the Heterogeneities at 34 °C

In a three dimensional object, the mean square displacement $\langle x^2 \rangle$ is related to the translational diffusion coefficient D_T and the characteristic translational diffusion time τ_T by

$$\langle x^2 \rangle = 6 D_T \tau_T. \tag{A.5}$$

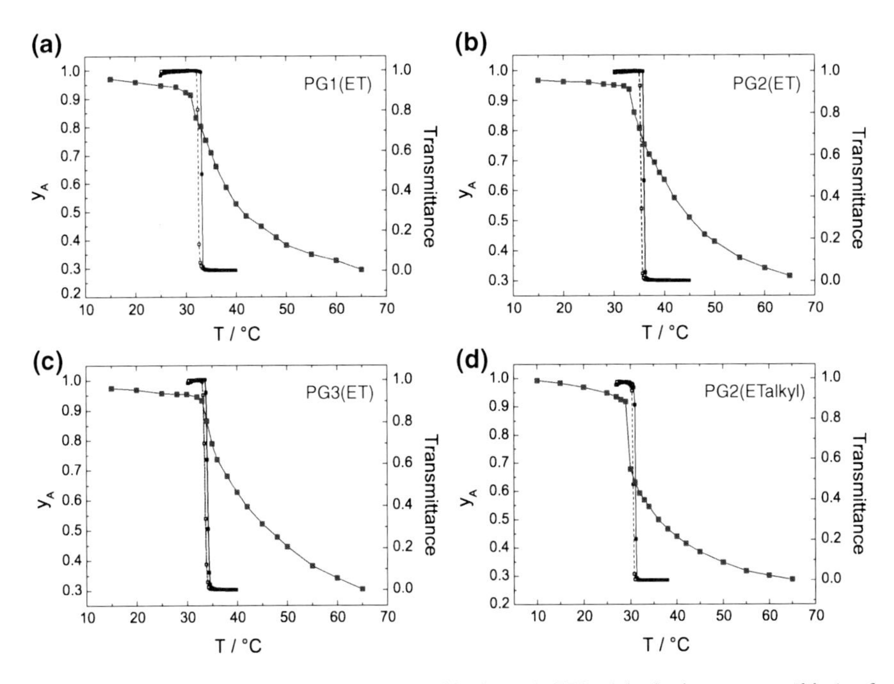

Fig. A.21 Comparison of turbidity curves (*black*) and EPR dehydration curves (*blue*) of **a** PG1(ET), **b** PG2(ET), **c** PG3(ET), and **d** PG2(ETalkyl). The turbidity curves were obtained for 0.25 wt% aqueous solutions of the dendronized polymers with a heating (*solid line*) and cooling (*dashed line*) rate of 0.2 K min^{-1}. The EPR data were obtained for 10 wt% aqueous polymer solutions by stepwise decreasing the temperature after ramp heating. Reprinted with permission from [18, 19]. Copyright 2010 WILEY-VCH Verlag GmbH & Co. KGaA, Weinheim

In the case at hand, the spin probe needs to diffuse fast enough from a hydrophobic into a hydrophilic region or vice versa, so that the dynamic exchange affects the CW EPR spectra and a motionally averaged line is observed. The additional breadth of this averaged line due to the dynamic exchange T_{exch}^{-1} is related to the spectral separation $\Delta\omega$ between the two EPR transitions and the reduced lifetime τ (Eq. 2.60) and is calculated by Eq. 2.61. The averaged line observed here has a breadth comparable to that of the two spectral components in the static limit ($T_{\mathrm{A/B}}^{-1}$ at 15 and 65 °C). Hence, no extra broadening of the line-width is observed due to chemical exchange and T_{exch}^{-1} can be estimated by

$$T_{\mathrm{exch}}^{-1} \leq \frac{1}{10} T_{\mathrm{A/B}}^{-1}. \tag{A.6}$$

For $T = 34$ °C, slightly above the critical temperature T_{C}, both spectral components contribute equally to the spectrum and Eq. 2.61 simplifies to

$$T_{\mathrm{exch}}^{-1} = \frac{1}{8}(\Delta\omega)^2\tau \leq \frac{1}{10} T_{\mathrm{A/B}}^{-1}. \tag{A.7}$$

With $\Delta\omega = 6.16$ MHz and $T_{A/B}^{-1} = 4.51$ MHz, the reduced lifetime can be estimated to $\tau \leq 95$ ns. In this framework, the characteristic translational diffusion time τ_T equals the reduced lifetime τ. With a rotational correlation time τ_c of 0.4 ns and $\tau_T = \tau \leq 240\,\tau_c$, Eq. A.4 can be rearranged to

$$\langle x^2 \rangle \leq 240 \cdot 6 D_T \tau_c. \tag{A.8}$$

Using the relation between the rotational correlation time and the rotational diffusion coefficient (Eq. 2.66), the mean square displacement is given by

$$\langle x^2 \rangle \leq 240 \frac{D_T}{D_R}. \tag{A.9}$$

Applying the Stokes–Einstein relation for a spherical particle with the hydrodynamic radius R_H, the diffusion coefficients are given by

$$\begin{aligned} D_T &= \frac{kT}{6\pi\eta R_{H,T}}, \\ D_R &= \frac{kT}{8\pi\eta R_{H,R}^3}, \end{aligned} \tag{A.10}$$

and the following relationship is obtained

$$\langle x^2 \rangle \leq 320 \frac{R_{H,R}^3}{R_{H,T}}. \tag{A.11}$$

$R_{H,R}$ and $R_{H,T}$ are the effective hydrodynamic radii for rotation and diffusion, respectively. Though both $R_{H,R}$ and $R_{H,T}$ strongly depend on the solvent properties, the ratio $R_{H,R}^3/R_{H,T}$ was found to be constant in solvents of different viscosity. For TEMPO, a value of $R_{H,R}^3/R_{H,T} = 0.08\ \text{nm}^2$ is reported [11].

Using this value, the mean square displacement amounts to $\langle x^2 \rangle \leq 25.6\ \text{nm}^2$, and $\langle x^2 \rangle^{1/2} \leq 5.1$ nm is obtained. This value is an upper limit estimate of the displacement due to the diffusion of the nitroxide. Though this diffusion will be assisted by fluctuations of the polymer undergoing the thermal transition, one can conclude that the size of inhomogeneities formed during the thermal transition is in the range of a few nm.

A.5 References

1. Lee TD, Keana FW (1975) J Org Chem 40:3145–3147
2. Paleos CM, Dais P (1977) J Chem Soc Chem Commun 345–346
3. Curry S, Mandelkow H, Brick P, Franks N (1998) Nat Struct Biol 5:827–835
4. Curry S (2009) Drug Metab Pharmacokinet 24:342–357
5. Spector AA (1975) J Lipid Res 16:165–179
6. Milov AD, Maryasov AG, Tsvetkov YD (1998) Appl Magn Reson 15:107–143

7. Jeschke G (2002) ChemPhysChem 3:927–932
8. Bhattacharya AA, Grüne T, Curry S (2000) J Mol Biol 303:721–732
9. Simard JR, Zunszain PA, Hamilton JA, Curry S (2006) J Mol Biol 361:336–351
10. Slutter CE, Gromov I, Epel B, Pecht I, Richards JH, Goldfarb D (2001) J Am Chem Soc 123:5325–5336
11. Kovarskii AL, Wasserman AM, Buchachenko AL (1972) J Magn Reson 7:225–237
12. Junk MJN, Spiess HW, Hinderberger D (2010) Angew Chem 122:8937–8941
13. Junk MJN, Spiess HW, Hinderberger D (2010) Angew Chem Int Ed 49:8755–8759
14. Junk MJN, Spiess HW, Hinderberger D (2011) J Magn Reson 210:210–217
15. Junk MJN, Spiess HW, Hinderberger D (2011) Biophys J 100:2293–2301
16. Zhang J, Junk MJN, Luo J, Hinderberger D, Zhu XX (2010) Langmuir 26:13415–13421
17. Junk MJN, Jonas U, Hinderberger D (2008) Small 4:1485–1493
18. Junk MJN, Li W, Schlüter AD, Wegner G, Spiess HW, Zhang A, Hinderberger D (2010) Angew Chem 122:5818–5823
19. Junk MJN, Li W, Schlüter AD, Wegner G, Spiess HW, Zhang A, Hinderberger D (2010) Angew Chem Int Ed 49:5683–5687

Curriculum Vitae

Matthias J. N. Junk

Born on July 28, 1983 in Bernkastel-Kues, Germany.

Education

Since Aug. 2010	Junior Specialist at the Department of Chemical Engineering at the University of California, Santa Barbara (Prof. B. F. Chmelka)
June 2010–July 2010	Post-doctoral scholar at the Max Planck Institute for Polymer Research, Mainz, Germany
June 2007–May 2010	Dissertation in Chemistry: "*Assessing the Functional Structure of Molecular Transporters by EPR Spectroscopy*" under supervision of Prof. H. W. Spiess at the Max Planck Institute for Polymer Research, Mainz, Germany Dr. rer. nat. (*summa cum laude*) granted May 31, 2010
Oct. 2006–May 2007	Diploma Thesis: "*Struktur und Strukturierung von photovernetzbaren, thermoresponsiven Hydrogelsystemen*" under supervision of Dr. U. Jonas and Prof. W. Knoll at the Max Planck Institute for Polymer Research, Mainz, Germany
Sept. 2005–Feb. 2006	Research student at the Department for Polymer Science and Engineering, University of Massachusetts in Amherst, USA; Scientific thesis: "*Sulfur-Based Molecular Fortifiers*" under supervision of Prof. A. J. Lesser

M. J. N. Junk, *Assessing the Functional Structure of Molecular Transporters by EPR Spectroscopy*, Springer Theses, DOI: 10.1007/978-3-642-25135-1,

Oct. 2002–May 2007	Studies in chemistry (Diploma) at the Johannes Gutenberg-University, Mainz
Sept. 2001–June 2002	Alternative civilian service at the "Bundesverband Selbsthilfe Körperbehinderter—Bereich Mittelmosel"
Aug. 1993–June 2001	Nikolaus-von-Kues-Gymnasium, Bernkastel-Kues Abitur granted June 12, 2001

Awards

2010	Otto Hahn Medal of the Max Planck Society
2010	MAINZ PhD Award of the Graduate School of Excellence "Materials Science in Mainz" (MAINZ)

Scholarships

Since Jan. 2011	Feodor Lynen scholarship of the Alexander von Humboldt- Foundation
June 2010–Dec. 2010	Scholarship of the Max Planck Society
Feb. 2008–Jan. 2010	Chemiefonds scholarship of the Chemical Industry Fund within the German Chemical Industry Association (VCI)
Nov. 2007–May 2010	Member of the Graduate School of Excellence "Materials Science in Mainz" (MAINZ)
Sept. 2005–Feb. 2006	Scholarship for studies at the University of Massachusetts in Amherst, USA granted by the German Academic Exchange Service
Oct. 2002–Sept. 2004	Anniversary scholarship of the Chemical Industry Fund within the German Chemical Industry Association (VCI)

Publications

[1] A. J. Lesser, K. Calzia, M. Junk, Reinforcement of Epoxy Networks With Sulfur- and Carbon-Based Molecular Fortifiers, *Polym. Eng. Sci.* **2007**, *47*, 1569–1575.
[2] M. J. N. Junk, U. Jonas, D. Hinderberger, EPR Spectroscopy Reveals Nano-Inhomogeneities in the Structure and Reactivity of Thermoresponsive Hydrogels, *Small* **2008**, *4*, 1485–1493.
[3] I. Anac, A. Aulasevich, M. J. N. Junk, P. Jakubowicz, R. F. Roskamp, B. Menges, U. Jonas, W. Knoll, Optical Characterization of Co-Nonsolvency Effects in Thin Responsive PNIPAAm-Based Gel Layers Exposed to Ethanol/Water Mixtures, *Macromol. Chem. Phys.* **2010**, *211*, 1018–1025.
[4] J. Zhang, J. Luo, X. X. Zhu, M. J. N. Junk, D. Hinderberger, Molecular Pockets Derived from Cholic Acid as Chemosensors for Metal Ions, *Langmuir* **2010**, *26*, 2958–2962.
[5] M. J. N. Junk, R. Berger, U. Jonas, Atomic Force Spectroscopy of Thermoresponsive Photo-Cross-Linked Hydrogel Films, *Langmuir* **2010**, *26*, 7262–7269.
[6] B. C. Dollmann, M. J. N. Junk, M. Drechsler, H. W. Spiess, D. Hinderberger, K. Münnemann, Thermoresponsive, Spin-labeled Hydrogels as Separable DNP Polarizing Agents, *Phys. Chem. Chem. Phys.* **2010**, *12*, 5879–5882.
[7] M. J. N. Junk, I. Anac, B. Menges, U. Jonas, Analysis of Optical Gradient Profiles During Temperature- and Salt-Dependent Swelling of Thin Responsive Hydrogel Films, *Langmuir* **2010**, *26*, 12253–12259.
[8] M. J. N. Junk, W. Li, A. D. Schlüter, G. Wegner, H. W. Spiess, A. Zhang, D. Hinderberger, EPR Spectroscopic Characterization of Local Nanoscopic Heterogeneities during the Thermal Collapse of Thermoresponsive Dendronized Polymers, *Angew. Chem.* **2010**, *122*, 5818–5823, *Angew. Chem. Int. Ed.* **2010**, *49*, 5683–5687.
[9] J. Zhang, M. J. N. Junk, J. Luo, D. Hinderberger, X. X. Zhu, 1,2,3-Triazole-containing Molecular Pockets Derived from Cholic Acid: The Influence of Structure on Host–Guest Coordination Properties, *Langmuir* **2010**, *26*, 13415–13421.
[10] M. J. N. Junk, H. W. Spiess, D. Hinderberger, The Distribution of Fatty Acids Reveals the Functional Structure of Human Serum Albumin, *Angew. Chem.* **2010**, *122*, 8937–8941, *Angew. Chem. Int. Ed.* **2010**, *49*, 8755–8759.
[11] Y. Akdogan, M. J. N. Junk, D. Hinderberger, Effect of Ionic Liquids on the Solution Structure of Human Serum Albumin, *Biomacromolecules* **2011**, *12*, 1072–1079.
[12] M. J. N. Junk, H. W. Spiess, D. Hinderberger, DEER in Biological Multispin-Systems: A Case Study on the Fatty Acid Binding to Human Serum Albumin, *J. Magn. Reson.* **2011**, *210*, 210–217.

[13] M. J. N. Junk, W. Li, A. D. Schlüter, G. Wegner, H. W. Spiess, A. Zhang, D. Hinderberger, EPR Spectroscopy Provides a Molecular View on Thermoresponsive Dendronized Polymers Below the Critical Temperature, *Macromol. Chem. Phys.* **2011**, *212*, 1229–1235.

[14] M. J. N. Junk, H. W. Spiess, D. Hinderberger, Characterization of the Solution Structure of Human Serum Albumin Loaded with a Metal Porphyrin and Fatty Acids, *Biophys. J.* **2011**, *100*, 2293–2301.

[15] M. J. N. Junk, W. Li, A. D. Schlüter, G. Wegner, H. W. Spiess, A. Zhang, D. Hinderberger, Formation of a Mesoscopic Skin Barrier in Mesoglobules of Thermoresponsive Polymers, *J. Am. Chem. Soc.* **2011**, *133*, 10832–10838.